UNIVERSE TRAVELER GUIDEBOOK

------

* OUR UNIVERSE *
* MULTIPLE UNIVERSES *

- How Universes Are Created -

- How Universes End –

Experiment, Observation, and Analysis

THIRD EDITION
DECEMBER 13, 2018

JOHN GILBERT BEAN

johngilbertbean.wordpress.com

OUR UNIVERSE AND MULTI-UNIVERSES BEYOND

All rights reserved. No part of This Book may be copied, reproduced, stored, or distributed in any form or media, or in any manner whatsoever without prior written permission of John Gilbert Bean or Maureen Bean.

Publisher's Cataloging-in-Publication

Bean, John Gilbert, 1933 –

UNIVERSE TRAVELER GUIDEBOOK / Maureen Bean – ed.
    p. cm.
    Includes index
    1. General Physics   2. Quantum /Virtual Particle Physics
    3. Relativity (Physics)   4. Cosmology – Popular Works
    I. Title
ISBN-13: 978-1975902599
ISBN-10: 1975902599

Published by: John Gilbert Bean Publishing
                SAN: 253-0287

                951-265-1124

                Send comments to: JBeanMBean@msn.com

1ST EDITION AUGUST 13, 2017
2ND EDITION OCTOBER 13, 2017
3RD EDITION DECEMBER 13, 2018

CONTENTS

PREFACE.
DEDICATION.
ABOUT THE AUTHOR.

INTRODUCTION - OUR UNIVERSE IS ONE OF MANY IN MULTI-UNIVERSE SPACE.

1. TIMELINE AND INFORMATION ABOUT OUR UNIVERSE.
2. EINSTEIN'S EQUATION FOR THE EQUIVALENCE OF MASS m AND ENERGY E.
2-A. WORK, ENERGY, MASS, AND CONSERVATION LAWS.
3. MASS (IN KILOGRAMS) OF VARIOUS OBJECTS – NEWTON'S LAW OF GRAVITY.
4. VELOCITY OF LIGHT AND OTHER ELECTROMAGNETIC WAVES.
5. ELECTROMAGNETIC SPECTRUM OF PHOTONS.
6. COSMIC RAYS.
7. COSMOLOGICAL UNITS.
8. AGE AND DIAMETER OF UNIVERSE.
9. TEMPERATURE - IN DEGREES.
10. KELVIN (K) TEMPERATURE OF VARIOUS PHENOMENA.

11. ELEMENTARY QUANTUM PARTICLES – THE VERY TINY BUILDING BLOCKS OF ATOMS.
11-A. ARE UNIVERSES MADE OF ANTI-ELECTRONS AND ANTI-PROTONS POSSIBLE?
11-B. QUANTUM PARTICLES (SECTION 11) ARE INFINITE PROBABILITY WAVES THAT INSTANTLY SPREAD AND EXTEND THROUGHOUT ENTIRE UNIVERSE.
12. HEISENBERG'S UNCERTAINTY PRINCIPLE.
13. de BROGLIE WAVE CHARACTERISTICS OF ELEMENTARY QUANTUM PARTICLES.
14. QUANTUM PARTICLE PROBABILITY WAVES.
15. ENTANGLED QUANTUM PARTICLES.
16. VIRTUAL QUANTUM PARTICLES AND VIRTUAL QUANTUM ANTI-PARTICLES.
17. BOSONS MEDIATE INTERACTIONS BETWEEN QUANTUM PARTICLES.
18. DESCRIPTION AND RELATIVE STRENGTH OF BOSONS.
19. HADRONS (MESONS, BOSONS, AND QUARK-COMPOSITES) - DESCRIPTION.

CONTENTS (CONTINUED)

20. EVERY QUANTUM PARTICLE IS PERTURBATION OF RELATED BOSON FORCE FIELD THAT STRETCHES TO INFINITY IN ALL DIRECTIONS: COULOMB'S LAW.

21. PARTICLE INTERACTIONS: THE HEART OF PARTICLE PHYSICS.

22. WAYS THAT QUANTUM PARTICLES INTERACT.

23. QUANTUM PARTICLE INTERACTION NOTATION.

23-A. CREATING ENERGY OUT OF MATTER.

23-B. CREATING MATTER OUT OF ENERGY.

23-C. QUANTUM PARTICLE SCATTERING.

23-D. CHARGED PARTICLES INTERACT THROUGH GLUONS.

24. ENERGY IS BORROWED FROM "EMPTY SPACE" TO CREATE VIRTUAL QUANTUM PARTICLES.

24-A. COULOMB'S LAW DEMONSTRATES EXISTENCE OF VIRTUAL PARTICLES.

24-B. NEWTON'S EQUATIONS DEMONSTRATE EXISTENCE OF VIRTUAL PARTICLES.

24-C. TIME IS OF THE ESSENCE IN A UNIVERSE: KEY ASPECTS OF TIME IN A UNIVERSE AND EMPTY SPACE BEYOND.

24-D. CONVERSION OF MANY, MANY SMALL AMOUNTS OF MASS OF VIRTUAL QUANTUM PARTICLES INTO HUGE AMOUNT OF ENERGY CREATES UNIVERSE IN "EMPTY" SPACE.

25. BIG BANGS CREATED OUR UNIVERSE AND OTHERS OUT OF ENERGY BORROWED FROM VIRTUAL PARTICLE PAIRS IN "EMPTY SPACE" THAT ANNIHILATE INTO PURE ENERGY.

25-A. HOW MUCH BORROWED ENERGY IS REQUIRED TO MAKE A UNIVERSE)?

25-B. BIG BANG CREATED OUR UNIVERSE OUT OF ENERGY BORROWED FROM VIRTUAL QUANTUM PARTICLE PAIRS IN INFINITE EMPTY SPACE THAT EXISTED BEFORE THE UNIVERSE EXISTED.

25-C. ENERGY TO MAKE A BIG BANG IS BORROWED ENERGY, CREATED FROM VIRTUAL QUANTUM PARTICLE PAIR ANNIHILATION IN EMPTY SPACE THAT EXISTED BEFORE OUR UNIVERSE, LEAVING BEHIND A UNIVERSE THAT MUST EVENTUALLY REPAY THIS BORROWED ENERGY.

26. CALCULATION OF QUANTITY OF VIRTUAL QUANTUM PARTICLE PAIRS THAT ANNIHILATED AT EXACT INSTANT AND PLACE OF BIG BANG.

26-A. DO OTHER UNIVERSES EXIST?

26-B. PAYING BACK ENERGY BORROWED TO MAKE G BANG.

27. TIMELINE, TEMPERATURE, AND PRODUCTS OF BIG BANG THAT CREATED OUR UNIVERSE.

28. TYPES OF STARS.

29. DYING STARS ARE FACTORIES THAT CREATED ALL THE ATOMS IN A UNIVERSE FROM HYDROGEN CLOUDS.

30. BLACK HOLES AND QUASARS.

31. COSMIC MICROWAVE BACKGROUND RADIATION (CMBR) IS EVIDENCE OF OUR BIG BANG.

32. DARK MATTER: 21 PERCENT OF OUR UNIVERSE; DARK ENERGY: 72 PERCENT; ORDINARY MATTER: 7 PERCENT.

33. COMPOSITION OF DARK MATTER AND ITS ORIGIN.

34. DARK ENERGY (72 PERCENT OF OUR UNIVERSE) IS CAUSING OUR UNIVERSE TO EXPAND AT EVER-INCREASING RATE.

35. GALAXIES ARE MOVING AWAY FROM EACH OTHER AT INCREASING VELOCITY CREATING SPACE BETWEEN GALAXIES - BUT GALAXIES ARE NOT EXPANDING.

36. THE END OF OUR UNIVERSE.

37. BIG BANGS CREATE OTHER UNIVERSES TOO, BUT NOT VERY OFTEN.

EPILOGUE   BIBLIOGRAPHY   INDEX

PREFACE.

The purpose of this book is to provide those who seek key information about our Solar System, our Galaxy, other Galaxies, our Universe, and other Universes in Multi-Universe Space beyond our Universe, including How and Where Universes Begin, and How Universes End.

Many people already know a lot about our Solar System and that it has a star at its center named "Sun." Fortunately, we live full time in that solar system on a beautiful planet called "Earth" that is in a hospitable orbit around "Sun" along with 8 or 9 other planets. Each of the other planets in our solar system is either too close to Sun or too far away from Sun to be livable.

Some people on planet Earth experience the best of times, but many experience times when it would be nice to Jump-Ship and leave Planet Earth for a more hospitable destination.

This book does not for provide solutions for problems on Planet Earth. This book is for those who want to look far beyond the small planet Earth and its solar system and travel to other solar systems in our Galaxy the Milky Way, to other Galaxies in Our Universe, and to other Universes in Multi-Universe Space beyond our Universe.

Bon Voyage!

JGB

DEDICATION.

This work is dedicated to the Late-Great Stephen Hawking who passed away recently. His theory of Black Hole radiation gave me a basis to understand the Creation and End of Universes and to extrapolate far beyond his genius. Thus, it was possible for me to deduce and understand the creation of and end of our Universe and other Universes in Multi-Universe Space. To my knowledge, my reasoning and conclusions as set forth and explained in this book and have not published previously by others elsewhere.

JGB

ABOUT THE AUTHOR.

John Gilbert Bean has a BS Degree in Physics from the University of Arizona and graduate study in Physics and Electrical Engineering at the University of Arizona and California State University at Northridge (formerly San Fernando Valley State College).

As well as writing, his career spanned Engineering, Engineering Management, and Program Management responsible for design, development and integration of State-of-the-Art Aircraft and Avionics Systems at RCA, Hughes Aircraft, and Northrop-Grumman.

His career and interests enabled him to work at the forefront of technology in the rapidly changing relevant science and mathematics of the origin and destiny of our Universe and Space beyond. He has also written:

1. **The Universe, Space, and Beyond**

2. **Communicating Successfully with Everyone**

3. **Get Rich and Stay Rich**

4. **How Russia Elected Trump –
How Trump Sold America's Security to Russia**

INTRODUCTION – OUR UNIVERSE IS ONE OF MANY IN MULTI-UNIVERSE SPACE

When I began the update of this book, Universe Traveler Guidebook, I first intended that it would be just a minor effort. However, as I continued my research and study, I realized that Our Universe is only part of the Big Picture and it fits in a larger scheme of many more Universes in Multi-Universe Space. This larger scheme requires redefining and discussing the composition and make-up of Space beyond our Universe - a Space that contains many Universes.

This Book is about a **MULTI-UNIVERSE SPACE --- SPACE that is everywhere, UNENDING,** and contains many Universes:

* A MULTITUDE OF UNIVERSES HAVE EXISTED IN THIS MULTI-UNIVERSE SPACE IN THE PAST, MANY OF WHICH NO LONGER EXIST.

* OUR UNIVERSE AND OTHER UNIVERSES CURRENTLY EXIST IN THIS MULTI-UNIVERSE SPACE BUT ALL THESE WILL EVENTUALLY DISAPPEAR.

* ADDITIONAL UNIVERSES WILL BE CREATED IN THIS MULTI-UNIVERSE SPACE IN THE FUTURE.

One of my objectives in life has been to understand what existed before our Universe, how our Universe came about, how it works, how it will end, and what will exist after our Universe. In my studies of our Universe, I found academic and popular physics literature about the Universe contained qualitative and qualitative information that was highly speculative and difficult for me to accept.

For Example: Some scientists believe that Space and Time did not exist before the "Big Bang" created our Universe from "Nothing," and that Space and Time were created as our Universe expanded from a Big Bang. This scenario is difficult to understand - if taken literally, there would have been **NO PLACE FOR THE BIG BANG TO GO BANG AND NO ENERGY TO CREATE IT.**

From my research, study, analysis, and extrapolation, I have determined that there are perhaps billions and trillions and more Universes much like ours in empty Space and all have the same physics. Some scientists believe that some or all these Multi-Universes operate with as many as 10^{500} different physics than the one our Universe has. - I do not.

Some scientists (INCLUDING ME) believe that "Wormholes" perhaps in "Black Holes" may provide pathways within our Universe and between other Universes because our Universe is curved.

To resolve conflicts in viewpoints, at least in my mind, I wrote my initial book about our Universe in 2007. I also wrote that book because, in addition to academic training and study of physics of the Universe, writing and teaching help me solidify my understanding and test my conclusions.

My initial books about our Universe have been revised several times to update them based on new developments in cosmology and particle physics experiments from sources such as the Hubble Telescope and the collider at CERN. Nevertheless, I felt these earlier books, main-stream cosmology, and physics dogma did not tell the complete story about our Universe and Multi-Universes.

After many sleepless nights of studying, and pondering, and thanks to Stephen Hawking, I was finally able to put the **last piece of the puzzle in place as to What existed before Our Universe, How the Big Bang that created Our Universe came about, How Our Universe will end, and What will Exist after Our Universe**. Hence, this Book! This book, thus, contains description of our Universe and Other Universes from **pre-Big Bang, to Big Bang creating a Universe, to expansion of a Universe as galaxies move apart, and finally to the end and aftermath of a Universe.**

This book is supported to the extent of current knowledge of experiment, observation, and analysis, and extrapolates beyond these to describe our Universe and other Universes.

INTRODUCTION – OUR UNIVERSE IS ONE OF MANY IN MULTI-UNIVERSE SPACE (CONTINUED)

Some readers might challenge my extrapolation and conclusions as some speculation on my part is necessary. However, I have striven to make all information herein consistent with experimental results and cosmological observations.

To enable understanding of the birth and death of a Universe, this book provides the reader with a key knowledge of Quantum Particles, Virtual Quantum Particles, and their associated Antiparticles. Quantum Particles and Virtual Quantum Particles are smaller than atoms and range to the smallest photon of electromagnetic energy. Examples: proton: 1.7×10^{-15} m, electron: 1.5×10^{-18} m, gamma-ray photon (emitted by atom or quantum particle): 3×10^{-20} m, Planck Length: 1.6×10^{-35} m.

Many books explain the mathematics of Quantum Mechanics, Relativistic Physics, and Cosmology. Those used in preparation of this book are listed in the Bibliography. The Goal of this book is to describe our Universe and other universes in terms that enable detailed understanding without the use of complex mathematics.

My earlier book <u>The Universe, Space, and Beyond</u> (6th Edition dated October 13, 2016) is a highly recommended companion to this book which provides expanded information on the general physics, mathematics, and cosmology of our Universe. This book and others in the Bibliography provide a range of mathematical sophistication.

I believe my explanations in this Book of the Physics of the End of a Universe (based on expanded theories of the Late Stephen Hawking) and my extrapolation to the physics of a Big-Bang-Birth of a Universe are correct and will stand the test of time.

JGB

OUR UNIVERSE AND MULTI-UNIVERSE SPACE BEYOND

John Gilbert Bean
johngilbertbean.wordpress.com

Send comments to JBeanMBean@msn.com

OUR UNIVERSE AND MULTI-UNIVERSES BEYOND

1. TIMELINE AND INFORMATION ABOUT OUR UNIVERSE.

- Number of Universes in existence
    At least 1, probably many more
- Our Universe created
    13.8 billion years ago
- Number of galaxies in our Universe
    150 billion (probably more)
- Number of stars in our Universe
    30 thousand billion, billion (300 sextillion) (probably more)
- Earth and Solar System created
    4.7 billion years ago
- Intense bombardment of Earth by meteors ended
    4.5 billion years ago
- First single-cell life on Earth
    4 billion years ago
- First multi-cell life on Earth (fractals)
    580 million years ago
- First life on earth that could move around
    550 million years ago
- First mammals and dinosaurs
    230 million years ago
- Dinosaurs dominate all land inhabitants
    200 million years ago
- Dinosaurs and many other species became extinct when large meteor crashed into Earth
    65 million years ago
- Early human ancestors left trees and walked upright with stiff arched feet
    6 million years ago
- Early human ancestors knap (chip) stones to create tools with sharp points and edges
    3.3 million years ago [Nature May, 2015]
- Early human ancestors lost most of their body hair
    1 million years ago
- Earliest evidence (a tooth) of Humans on Earth (in central Israel not Africa)
    400,000 years ago
- Humans began wearing clothing
    170,000 years ago

OUR UNIVERSE AND MULTI-UNIVERSES BEYOND

- Humans migrated out of Africa
    - 125,000 years ago
- Neanderthals interbreed with Humans
    - 80,000 to 40.000 years ago
- Humans migrated from Asia to North America via Alaska
    - 15, 000 years ago
- Written Language invented
    - 10,000 years ago
- Archimedes developed early physics laws and calculus-like methods
    - 2265 years ago   250 BC
- Isaac Newton revolutionized Physics, Mathematics, and Engineering with his Calculus, Laws of Motion, and Theory of Gravity
    - 336 years ago   1679
- Charles Darwin's Theory of Evolution
    - 158 years ago   1859
- James Maxwell's Equations of Electromagnetic Fields
    - 153 years ago   1864
- Einstein's Special Theory of Relativity
    - 112 years ago   1905
- Niels Bohr developed one-electron model of atom
    - 104 years ago   1913
- Einstein's General Theory of Relativity
    - 120 years ago   1915
- Quantum Mechanics developed
    - 92 years ago   1925
- Edwin Hubble found the Universe is expanding.
    - 89 years ago   1928
- Niels Bohr and G. V. Wheeler explained neutron role in fission of Uranium 235
    - 78 years ago   1939
- Big Bang Theory of creation of Universe
    - 67 years ago   1950
- Standard Model of Elementary Particles
    - 47 years ago   1970
- Inflationary Theory of Creation of the Universe   47 years ago   1970

1. TIMELINE AND INFORMATION ABOUT OUR UNIVERSE (CONTINUED).
- Higgs boson detected to a certainty at CERN
    5 years ago   2012

Super-symmetric particles being searched for at CERN
    2013 and on going

Experimental evidence of inflation of our Universe at origin found in CMBR polarization
    3 years ago   2014

- Stephen Hawking's Theory of Black Hole Radiation: Black Holes radiate virtual particles and will eventually disappear
    1 year ago   2017

- String Theory/M-Theory of Elementary Particles (Unified Theory of Everything)
    Under development

- Dark Matter and Dark Energy
    Being searched for extensively

The Sun will exhaust its fuel, greatly expand, and burn up the Earth
    7 Billion years from now

The Universe will be completely cold, radiate away, and return from whence it came
    A few hundred billion years or so from now

2. EINSTEIN'S EQUATION FOR THE EQUIVALENCE OF MASS m AND ENERGY E.

$E = mc^2$ Where c is the velocity of light = $2.99\ 792\ 458 \times 10^8$ m/sec^{-1}

This equation shows that mass and energy are the same thing, it takes only a little tiny bit of mass to make a lot of energy, but it takes lots and lots of energy to make a little tiny bit of mass.

The energy it takes to create mass is exactly balanced by the energy of the mass that has been created, and vice versa. This principle allows the creation of a Universe. Both mass and energy have gravitational attraction. An object weights more when it is hot (perhaps exploding) than cold – when hotter, it has more energy, thus "more gravity." Clocks run slower in a high-strength gravitational field and faster in a weaker gravitational field according to Einstein's General and Special Relativity Theories.

2-A. WORK, ENERGY, MASS, AND CONSERVATION LAWS.

Work (**W**) on an object is done by a Force (F) moving an object (mass) a distance (**s**) with an acceleration (**a**). Work = Force F times distance s.

Energy is the ability to do Work. The two types of energy are kinetic (motion) energy (such a falling rock) and potential (static at rest) energy (such as a car stopped at the top of a hill). Forces (Section 3) involved with energy are conservative force (without energy loss) and dissipative force (with loss such as due to friction).

Energy and Work have the same units (kg-m^2/sec^2). **Energy** can be positive or negative. **Pressure** can also do work and be positive (outward) or negative (vacuum).

Kinds of energy are: Gravitational (Section 3), Thermal (Sections 9 and 10), Chemical/Nuclear (Section 21), Mass (Section 2), Radiant (Section 5), Mechanical.

The principle of conservation of total energy is used in explanations in many places in this book. **It is a key principle of all physics and all chemistry.** Total energy E is the sum of kinetic (motion) K and potential (static) U energies. Involved. Total energy is always conserved less any energy dissipated by friction, heat, or otherwise. **Positive (or negative) Energy can be borrowed, but must eventually be paid back with Negative (or positive) Energy, such as with a swinging pendulum.**

A Universe cannot be created out of nothing! **Positive (or Negative) Energy can be borrowed from Virtual Quantum Particles and Virtual Quantum Antiparticles in Empty Space to create a Positive (or Negative) Universe in Empty Space, as discussed later (Section 16), but the borrowing leaves behind Negative (or Positive) Energy in Empty Space which EVENTUALLY MUST EVENTUALLY BE FULLY PAID BACK by the return of the borrowed Positive (or Negative) Energy.**

Drop any object. Watch as the potential energy of the object due to gravity is converted to kinetic energy of motion as the DROPPED object borrows energy from gravity, accelerates, falls toward the floor or the ground - and then repays the *borrowed* energy as it hits the ground. Did you notice that the Earth fell toward the falling object? It did, but the Earth is so massive its motion was imperceptible. Nevertheless, when the object was falling, the Earth was falling very slightly toward it. Then when the object hit the Earth, the Earth recoiled slightly, but not totally toward its original position.

There are many Conservation Laws in the physics of interactions of our Earth, Galaxy, and Universe. Many more will be encountered in the paragraph below and throughout the rest of this book. Application of these laws explains both the creation and end of our Universe and other universes. **Examples of Conservation Laws in the Physics of our Universe are Charge (+ and -), momentum (mass times velocity), heat, mass, spin (up and down), color (red, green, and blue), etc. In every interaction, conservation laws are verified. For example if a negative (–) particle interacts with a positive (+) particle, the two charges cancel and the resultant particle must be charge neutral. Any discrepancy must be resolved, and resolving the discrepancies has led to many, many advances in science.**

3. MASS (IN KILOGRAMS) OF VARIOUS OBJECTS - NEWTON'S LAW OF GRAVITY.

Electron	= 9.109×10^{-31}	Proton	= 1.673×10^{-27}
Oxygen atom	= 3×10^{-26}	Ant	= 10^{-5}
Humming bird	= 10^{-2}	Man (220 pounds)	= 10^2
Elephant	= 10^4	Whale	= 10^5
Ship	= 10^8	Moon	= 7×10^{22}
Earth	= 5.972×10^{24}	Sun	= 1.988×10^{30}
Milky Way Galaxy	= 2×10^{41}	Universe	= 3×10^{56}

NOTE: The term "mass" as used in this book usually means the composite of rest mass m_o, relativistic mass m_r and energy mass m_e. Relativistic mass and energy mass must be considered when extreme circumstances apply such as of velocities approaching the velocity of light and/or large objects such as the Sun, galaxy, or black hole.

The Force of Attraction of Masses m_1 and m_2 a distance r apart is Newton's Law of Gravity:

$$F_g = G(m_1 m_2)/r^2$$

G is the Gravitational Constant = 6.67×10^{-11} N m² kg⁻²

Combining the information in this section about the force of gravity F_g with Einstein's law of the equivalence of mass and energy (in Section 2) leads to an explanation of "Gravity Waves." Any sudden source of energy such as a firecracker, a stick of dynamite, or explosion of a star as a supernova (Section 28) will cause a surge in its gravity. The change in energy in most explosions is too small to be significant.

Einstein predicted that the explosion of a distant giant supernova could result in a "Gravity Wave" that could be detected on Earth and would be a confirmation of his Equation given in his General theory of Relativity. Such gravity waves were detected in March, 2018.

4. VELOCITY OF LIGHT AND OTHER ELECTROMAGNETIC WAVES.

    186,000 miles per second
    299,792,458 meters per second = 3×10^8 m/sec

5. ELECTROMAGNETIC SPECTRUM OF PHOTONS.
Frequency is measured in Hertz (Hz). 1 Hz – 1 cycle per second (cps)

Household 60 Hz wall power λ = 3100 miles
AM radio 1000 KHz λ = 982 ft (299 meters)
Channel 2 TV (54 MHz) λ = 18.19 ft (5.54 meters)
Microwave (10^{11} Hz) λ = 2.99×10^{-3} meters = 2.99 mm
Visible light (10^{14} Hz) λ = 2.99×10^{-6} meters
X-ray (10^{18} Hz) λ = 2.99×10^{-10} meters
Gamma rays (10^{21} Hz) λ = 2.99×10^{-13} meters
Angstrom: $\dot{A} \equiv$ 0.1 nano-meter = 1×10^{-10} meters
The Angstrom (named after Swedish astronomer Anders Angstrom) is also used to express electromagnetic wavelengths of spectroscopy.

6. COSMIC RAYS.

Cosmic rays are particles from the sun and other solar systems that bombard the Earth. Cosmic rays have very short wavelengths and enormous energy mainly from very high velocity protons.

The radiation is 90 percent protons and 10 percent helium nuclei. Most of these are dissipated in the upper atmosphere. Pions carry away about half the incident energy. Collisions and decays result producing muons, electrons, neutrinos, and photons.

Cosmic rays are also used at high altitude laboratories or in space for experiments of high-energy particle interactions.

My first introduction to cosmic rays was at the lab at the top of the Physics Building at the University of Colorado in Boulder, Colorado
 - Elevation 6250 feet.

Energies of cosmic-ray particles are usually between 10 MeV and 10 GeV (10^7 eV and 10^{10} eV). Cosmic rays of energies of 10^{20} Gev have been detected.

7. COSMOLOGICAL UNITS.

Velocity of light $\quad c = 2.99\,792\,458 \times 10^8$ m/sec
Newtonian Gravitational Constant $\quad G = G_N = 6.6739 \times 10^{-11}$ m^3 kg^{-1} s^{-2}
Light year (ly) $\quad 0.946053 \times 10^{16}$ m
Astronomical Unit (au or A) $\quad 1.495\,978\,707\,00 \times 10^{11}$ m
Parsec (pc) $\quad 3.0856776 \times 10^{16}$ m = 3.262 ly = 2.06267×10^5 A

First Galaxies form $\quad$ 800 years after Big Bang
Galaxies in Universe $\quad 4 \times 10^{11}$
Stars per galaxy $\quad 1 \times 10^{12}$
Stars in Universe $\quad 4 \times 10^{23}$
New stars now being created by typical Starburst galaxy $\quad$ 2000 per year
New stars now being created by Milky Way galaxy $\quad$ 2 per year

Diameter of Milky Way $\quad$ 6 trillion miles = 6×10^{12} miles = 9.6×10^{12} km

Mass of Universe $\quad 3 \times 10^{56}$ kg
Energy of Universe $\quad 10^{50}$ kg = 9×10^{66} joules
Density of Universe $\quad 10^{-23}$ gm/m^3

Age of Universe $\quad 13.8 \times 10^9$ years
Radius of Universe $\quad 25 \times 10^9$ light-years = 25×10^9 (0.946053×10^{16} m)
$\quad\quad\quad\quad\quad\quad\quad = (2.365 \times 10^{26}$ m

Volume of Universe $\quad = (4/3)\,(\pi)\,(r^3) = (4/3)\,(3.14159)\,(6.355314 \times 10^{24}$ m)
$\quad\quad\quad\quad\quad\quad\quad = 1075 \times 10^{72}$ m^3

OUR UNIVERSE AND MULTI-UNIVERSES BEYOND

CMBR temperature   2.725 K   (Cosmic Microwave Background Radiation)
Stars in Milky Way Galaxy   300 billion = 3×10^{11}
Mass of Milky Way Galaxy 2×10^{41} kg
Planets in Milky Way Galaxy   50 billion

Our Solar System, its Planets, and Pluto formed   4.6 billion years ago
Radius of Sun 6.9551×10^8 m
Mass of Sun (Msolar) = 1 Solar mass = 1.9885×10^{30} kg
Sun Core Temperature 1.5×10^7 K
Sun Surface Temperature 6×10^3 K
Luminosity of Sun (LSun) 3.828×10^{26} W (watts)

Mean Distance from Sun to Earth   149,597,870,700 m
Mass of Earth 5.9727×10^{24} kg
Radius of Earth 6.378137×10^6 m

In contrast, here are some units used frequently on Earth:

1 teaspoon (tsp) = 6.0 grams (salt)   1 Tablespoon (tbsp) = 3 tsp
1 Cup = 8 fluid ounces = 16 Tablespoons = 48 teaspoons = 237 milliliters
1 C = 8 oz = 16 Tbsp = 48 tsp = 237 ml

K = 10^3;   M = 10^6;   G = 10^9;   T = 10^{12}

8. AGE AND DIAMETER OF OUR UNIVERSE.

Age of the Universe = 13.8 billion years

The Diameter of the Universe just before the BIG BANG when the Universe was a tiny sphere of hot energy of Virtual Quantum Particles crushed by four unified quantum forces (Section 25):

10^{-10} centimeter - or perhaps as small as 10^{-33} centimeters

Current Diameter of Universe:
40 to 90 billion light years (and expanding rapidly) assuming the Earth is about in the center of the Universe.

9. TEMPERATURE - IN DEGREES.
Centigrade C = 5/9(F -32)   F = Fahrenheit   Kelvin K = C + 273.15

Absolute zero temperature is the lowest possible obtainable temperature.
The lowest temperature achieved experimentally is 0.003 K.

As temperature decreases, elements change phase from Gas, to Liquid, to Solid, to Bose-Einstein Condensate. At temperatures below 2 K, startling effects occur, including superconductivity (zero resistance to electrical current).

Temperature is directly related to ability of Heat Energy to do Work and Kinetic Energy. The hotter (higher temperature) of an object, the more rapidly can heat energy transfer from a hotter to a colder object.

OUR UNIVERSE AND MULTI-UNIVERSES BEYOND

K^0

10. KELVIN (K) TEMPERATURE OF VARIOUS PHENOMENA.

ABSOLUTE ZERO K   0   (-273.15 C)   (- 459.665 F)

Planned Lab Experiment: 0.000,000,000,1   (1 ten-billionth of a degree above absolute zero K)

Lowest Temperature Achieved 0.003   (- 273.147 C) In a laboratory in Europe 12/2017

Elements Act Like Waves   1   Rather than like particles and can join other waves to make big waves

Temperature of Empty Space   2 (?)   Beyond our Universe

Bose-Einstein Condensate   2 Forms with all elements in lowest Energy States.

2.75 CMBR Background radiation

Temperature of Empty Space   3   In our Universe between galaxies

Helium Liquefies   4

Dry Ice (CO_2) Freezes   195

Water Freezes (solid)   273.15   0 C   (Zero degrees Centigrade)

Human Body Temperature   310

Water Boils (gas)   373.15   100 C

Gold Melts   1336

Surface of Sun 6×10^3

Center of Earth 1.6×10^4

Center of Sun 1.5×10^8

Center of a Giant Star 1×10^{10} (20 times bigger than Sun)

In December 2017, Researchers in Switzerland and Germany achieved the lowest temperature to date = 0.003 K

Planned future lab experiments will attempt to reach within one ten billionth (1/10, 000,000,000) of a degree above absolute zero K.

11. ELEMENTARY QUANTUM PARTICLES – THE VERY TINY BUILDING BLOCKS OF ATOMS.

39 ELEMENTARY PARTICLES: FERMIONS (LEPTONS AND QUARKS) AND BOSONS. LEPTONS CARRY CHARGE OF 0 or -1. QUARKS CARRY CHARGE OF +2/3 OR -1/3:

- 3 ELECTROMAGNETICALLY NEUTRAL LEPTONS (NEUTRINOS): ELECTRON NEUTRINO, MUON NEUTRINO, & TAU NEUTRINO.

- 3 ELECTROMAGNETICALLY CHARGED LEPTONS: ELECTRON, MUON, & TAU.

- 18 QUARKS: UP, CHARMED, TOP, DOWN, STRANGE, BOTTOM: EACH OF RED, BLUE, OR GREEN.

- 15 BOSONS: EACH BOSON IS ITS OWN ANTI-BOSON.
    - 1 ELECTROMAGNETIC PHOTON.
    - 2 WEAK NUCLEAR FORCE BOSONS W^+ (W^-), Z^0
    - 8 STRONG FORCE GLUONS.
    - 4 GRAVITATIONAL FORCE BOSONS (H^0 H^+ H^- GRAVITON).

285 COMPOSITE PARTICLES – HADRONS (MESONS AND BARYONS)
- 197 MESONS
- TETRA-QUARK, PENTA-QUARK
- 86 BARYONS.

Quantities are as reported by Collider at CERN in 2014. UNDETECTED QUARK COMBINATIONS IN MORE COMPLEX COMBINATIONS MAY BE ORIGIN OF DARK MATTER.

NOTE. There are 18 Quarks. Each of 6 different quarks (up u, down d, charmed, strange, top, bottom) can have the color red, blue or green making 18 in all. Each quark has a distinctive anti-quark. Free quarks cannot exist, but must be combined into a hadron (meson or baryon). Mesons consist of a quark and an antiquark. Baryons consist of three quarks or three antiquarks.

Mesons and baryons are colorless and have canceling colors. A colorless red meson can only be formed with a red quark and an anti-red quark. A colorless baryon is formed with 3 quarks each of a different color: red, green, and blue.

There are 197 mesons and 86 baryons. (More will be detected.) Only quarks in the **proton** baryon and **neutron** baryon exist for more than a millionth of a second. Thus, only the particles listed below behave as elementary particles.

- a. n Neutron (udd) Lifetime = **880.1 seconds (14.668 minutes)**
  (Unless securely bound to a proton inside an atom)

- b. p Proton (uud) Lifetime = ∞ (Infinity) **mass = 1.7 x 10^{-24} grams = 938 Mev/c^2**

- c. e Electron Lifetime = ∞ **mass = 9.1 x 10^{-28} grams = 0.51 Mev/c^2**

- d. ɣ Neutrinos (3) Lifetime = ∞
  - $ɣ_e$ Electron neutrino **mass = 1 x 10^{-6} Mev/c^2**
  - $ɣ_μ$ Muon neutrino **mass = 2 x 10^{-6} Mev/c^2**
  - $ɣ_T$ Tau neutrino **mass = 3 x 10^{-6} Mev/c^2**

Mass of a quantum particle is stated in grams and MeV/c^2 : Million electron Volts/c^2
  1. 2 Mev/c^2 = 18.2 x 10^{-28} grams

- e. ϒ Photon mass = 0 Lifetime = ∞   6. G Gluons (8) mass = 0 Lifetime = ∞

The proton, neutron, and electron form all the atoms that make up all of the material our Universe. All the rest of the particles are unstable and quickly decay into other unstable particles which then decay into other unstable particles, and so forth, until they decay into one or more of the above stable particles. Under all circumstances, the total energy, charge, and other parameters are conserved (the same before and after the decays). Each elementary particle is theorized to have a corresponding Super-Symmetric Particle. Super-Symmetric Particles have not been found to exist but the search is on. Each of the above 39 elementary particles has an antiparticle or is its own antiparticle that can exactly annihilate it into pure energy and/or other particles. Fortunately, when our Universe was created, there was a propensity to create more electrons (-) and protons (+) with lifetimes approaching infinity rather than antiparticles: anti-electrons (+) and anti-protons (-). So our Universe is made essentially of electrons and protons.

11-A. ARE UNIVERSES MADE OF ANTI-ELECTRONS AND ANTI-PROTONS POSSIBLE?

From the previous Section, the question arises, "Is a Universe made of quantum antiparticles which have the polarities of their charges and other characteristics of particles of our Universe reversed possible? If so, then elements there would be made of anti-electrons e^+ and anti-protons p^- . **An anti-electron is also called the positron. An antiproton p^- is also called the p bar or proton-bar.**

Anti-electrons e^+ and anti-protons p^- have been made in the by physicists in many laboratories in many experiments but the antiparticles and the anti-elements fabricated from them are usually short-lived and quickly annihilated.

No one knows for certain if an Anti-Universe is possible. The Reader can find extensive additional speculation about negative Universes in the literature.

Nevertheless, To-This-Writer, based on the current state of knowledge, it seems very likely, because Universes are so complicated and elegant in their physics, that only positive Universes of infinite-lived electrons and infinite-lived protons (Section 11) can be created and that a negative Universe would quickly destroy itself very early in its evolution. Nature seems too conservative to waste creation on Universes that would destroy each other and everything in them.

11-B. QUANTUM PARTICLES (SECTION 11) ARE INFINITE PROBABILITY WAVES THAT INSTANTLY SPREAD AND EXTEND THROUGHOUT ENTIRE UNIVERSE.

A Quantum Particle (Section 11) consists of probability waves that spread and exist throughout the UNIVERSE. It does not exist at a specific location until it is forced to select a specific location such as in an atom or to be annihilated by its Quantum AntiParticle. The probability of any individual free quantum particle existing each of its possible specific locations can be accurately calculated. The accuracy of the probability calculations by physicists is unprecedented. The location of any one particle cannot be determined, only the probability of it being at any of is possible specific locations. Thus, quantum particles behave as probability waves that stretch everywhere to infinity. The probability total of all the possible locations of any free quantum particle is 1.

Probabilities rule at the quantum level where if something can happen, it will, but maybe not very often. It is impossible to tell in advance which of billions and billions of possible paths a single particle such as a quantum photon will take. But, its most probable locations and paths can be calculated. It is impossible to tell in advance the path or location for any individual quantum particle, which can select any of its perhaps billions possible paths anywhere in the Universe. However, over billions of trials, the paths chosen by the quantum particles will always conform to the calculated possible probabilities for each possible path and location.

Physicists are able to calculate expected experimental results using the probability of a particle selecting a specific path, having certain characteristics and being created and existing at a certain place at a certain time – the many calculations required can be carried out to a very high degree of accuracy (10 or more decimal places). The sum of all the different probabilities is 1. This enables expected results of experiments to be accurately calculated in advance to as many decimal places as needed – a feat unprecedented in other sciences.

When a quantum particle is created, rather than existing in one place like a baseball, it is instantly created as a "probability wave" **stretching to infinity** – a separate probability wave to infinity for each of its possible path and locations. When for one reason or another, one of its possible paths is selected, all the other possible paths collapse instantly everywhere. See Section 20 for additional information.

12. HEISENBERG'S UNCERTAINTY PRINCIPLE.

Heisenberg's Uncertainty Principle states that at the quantum (tiny) particle level, if we want to "know" various characteristics of a quantum particle or quantum energy, and try to measure them, our first measurement to determine one of the characteristics will disturb all the other values. Thus, we can never know with certainty all information about a quantum particle or energy. For example, if we want to know the location and velocity of a quantum particle or energy, if we measure one aspect, we will disturb and change the other(s).

13. de BROGLIE WAVE CHARACTERISTICS OF ELEMENTARY QUANTUM PARTICLES.

Every object, large and small in the Universe exhibits particle and wave characteristics with a de Broglie wavelength. Elementary particles, hadrons, atoms, molecules, and even large objects exhibit both particle and de Broglie wave characteristics.

The wave characteristics of an object are important primarily for small quantum particles. Look at a distant light source through a small slit between two fingers on one hand and see the thin black lines between your fingers made as light photons interfere with each other due to their wave characteristics. There are many experiments verifying the dual particle and wave characteristics of quantum particles and large objects where coherent particle waves interfere with each other to produce interference patterns. For detailed information, refer to <u>Feynman</u> in the References or any general physics book.

14. QUANTUM PARTICLE PROBABILITY WAVES. The discussion above, about quantum particles and quantum antiparticles, explains that any single quantum particle or quantum antiparticle literally exists <u>everywhere</u> at once, leads to the premise that each of these quantum particles is composed of billions and billions of smaller components commonly called Strings in advanced mathematics. These "tiny round or flat strings" or "tiny-diameter round or rectangular small rods" are postulated to vibrate in multiple (usually ten) dimensions much like a violin string vibrates in four dimensions (three plus time). Each quantum particle has its own frequencies, harmonics, phases, and modes of vibration (as usually described in college physics books for large objects). A very extensive and expanded mathematical treatment of limited usefulness called "String Theory" has been formulated. Many other expanded and alternative theories have been developed: "SuperStrings," "Symmetry," "Super Symmetry," "Theory of Everything," and "Energy of Emptiness of Space" also with limited success to make predictions about our Universe and beyond. This last, the "Energy of Emptiness of Space" is finally on a track (although circuitous) to the "Theory of the Universe and Space Beyond" developed by this author, for this book, and documented throughout the sections of this book:

> The "Universe" and "Empty Space Beyond" are always Energy neutral and always contain equal parts of Negative and Positive Energy. They are a source of unlimited POSITIVE ENERGY and unlimited NEGATIVE ENERGY which can be "Borrowed in equal amounts to spontaneously create <u>Virtual</u> Quantum Particle Positive and Negative pairs. Positive and Negative Energy can come into existence by being "Borrowed in equal amounts" but are always eventually returned to the Universe or Empty Space from which they were borrowed by canceling Negative and Positive Virtual Quantum Pairs.

This premise leads to a simple, straight-forward explanations of the origin and end of a Universe in subsequent Sections, unlike the extremely complex theories listed at the beginning of this Section. As Einstein said, "The simplest explanation is usually the correct one. It is expected that the theory herein will survive the Test of Time and Review.

15. ENTANGLED QUANTUM PARTICLES.

Some quantum particles are created as entangled pairs even if each one of the pair is billions of kilometers or more (infinity) apart from the other of the pair. Certain of their characteristics such as "spin" may be "entangled." If one particle is found to have clockwise spin, then the other entangled particle has to have the same spin (or in some cases opposite spin) even if the selection of spin (clockwise or counterclockwise) is arbitrary for the first particle. This is the case even if there is "seemingly no other "connection" of any type between the two entangled particles and the entangled particles are great distances (approaching infinity) apart. It's like throwing a die on Earth which then shows a random number and finding its entangled partner billions of miles away always has the same number.

Further, the entanglement between the two entangled particles is instantaneous (travels faster than the speed of light) – which is seemingly impossible. If the two locations have synchronized clocks, and the time the spin (or other entangled characteristic of the two particles) is noted when the value of the characteristic is set for one entangled particle, the characteristic of the second entangled particle will be set the same **at exactly the same time** instantly faster than the speed of light at huge distances away to the same arbitrary value as the other particle's characteristic. This phenomenon is due quantum particles having the **characteristics of an infinite field**. **EVERY QUANTUM PARTICLE IS is not a localized entity but a PERTURBATION OF ITS RELATED BOSON FORCE FIELD THAT STRETCHES TO INFINITY IN ALL DIRECTIONS. SEE COULOMB'S LAW IN SECTION 20.**

Current communication systems are limited to speed of light. However, entangled particles, thus, provide a potential way for instant communication over trillions of miles faster than the speed of light. Research is now being done to attempt to build this type of communication system.

IF PATHS OR CHARACTERISTICS (STATES) OF TWO OR MORE ENTANGLED PARTICLES ARE MANDATED OR PROHIBITED, ALLOWED PATHS OR STATES ARE INSTANTLY SELECTED (FASTER THAN THE VELOCITY OF LIGHT) EVERYWHERE IN THE UNIVERSE WHEN ARBITRARILY SELECTED BY EITHER ONE OF THE ENTANGLED PARTICLES.

16. VIRTUAL QUANTUM PARTICLES AND VIRTUAL QUANTUM ANTI-PARTICLES.

EMPTY SPACE EVERYWHERE (IN UNIVERSES AND IN EMPTY SPACE BEYOND UNIVERSES) IS NOT EMPTY BUT IS FILLED WITH BILLIONS AND BILLIONS AND BILLIONS AND MORE OF "VIRTUAL" QUANTUM PARTICLE AND "VIRTUAL" QUANTUM ANTIPARTICLE PAIRS CREATED CONTINUOUSLY BY BORROWING Positive and Negative ENERGY FROM "EMPTY SPACE" OF THE UNIVERSE.

The term "Virtual" applies to Quantum Particles that are created spontaneously as "Virtual" Quantum Particle and Virtual Anti-Quantum Particle Pairs BY BORROWING ENERGY FROM A UNIVERSE OR EMPTY SPACE BEYOND A UNIVERSE.

The 39 elementary quantum particles and their quantum antiparticles, described in *Section 11*, are created **everywhere spontaneously as quantum pairs of virtual particles and their virtual quantum antiparticles.** These virtual particles and virtual antiparticles are always everywhere. They are SPONTANEOUSLY and CONTINUOUSLY CREATED EVERYWHERE AND OCCUR EVERYWHERE IN OUR UNIVERSE, IN NEWLY CREATED SPACE BETWEEN OUR GALAXIES AS GALAXIES MOVE APART WITHIN OUR UNIVERSE, AND IN EMPTY SPACE BEYOND OUR UNIVERSE.

IT TAKES ENERGY TO CREATE VIRTUAL QUANTUM PARTICLE PAIRS: ENERGY "BORROWED" FROM "EMPTY SPACE" OR A UNIVERSE (SEE SECTION 24.)
AN INFINITE AMOUNT OF VIRTUAL ENERGY MAY BE BORROWED TO CREATE VIRTUAL QUANTUM PARTICLE PAIRS IN OUR UNIVERSE AND IN EMPTY SPACE BEYOND OUR UNIVERSE.
<u>*BUT THE BORROWED ENERGY MUST BE PAID BACK.*</u> THE BORROWED ENERGY WILL ULTIMATELY WILL BE PAID WHEN THESE QUANTUM PARTICLES OF ENERGY COME TOGETHER LATER WITH THEIR ANTI-PARTICLE. THEN THE BORROWED ENERGY PAIR WILL ANNIHILATE INTO PURE ENERGY (AND PERHAPS OTHER PARTICLES). **(See Section 24.)**

ALL BORROWED ENERGY MUST EVENTUALLY BE FULLY PAID BACK!

EMPTY SPACE BEYOND OUR UNIVERSE IS POSTULATED IN THIS BOOK TO EXIST AND CONTAIN VIRTUAL QUANTUM PARTICLE PAIRS THAT ARE CONSTANTLY BEING CREATED AND LATER BEING ANNIHILATED BY THEIR ANTIPARTICLE OPPOSITES.

Empty space is created as the galaxies in our Universe move apart at an increasing velocity. This "empty space" is also postulated herein to contain virtual quantum particles. **This postulate may be demonstrated and confirmed by the Voyager Space Program which will send a spacecraft beyond our galaxy in the near future. While writing this book, the Voyager Program Office was contacted to suggest such an experiment.** In our Universe and beyond our Universe, a virtual quantum particle and its companion virtual quantum antiparticle may be created concurrently very close together or an enormous distance, perhaps billions and trillions of kilometers or more apart.

A virtual quantum particle is usually quickly annihilated by its closest virtual quantum antiparticle into pure energy, and may further decay spontaneously into another particle(s**). See Section 23**. It does not have to combine with the particle or anti-virtual particle it was paired with at its creation unless it was created entangled with the other particle. It is impossible to tell a virtual particle from a particle of the same kind, or a virtual-antiparticle from an antiparticle of the same kind. An electron is exactly identical with every other electron and virtual electron. A newly created virtual electron is exactly identical with every other electron and virtual electron. Any quantum particle is exactly the identical to every other particle of the same kind. As noted above, at creation, however, some particles may be "entangled" with another particle.

- Without virtual particles and virtual antiparticles, there would have been no Big Bang (as described later) and neither our Universe nor any other universe could ever exist.

- Without virtual particles and virtual antiparticles in being created in Black Holes (to be described later), the end of our Universe would be drastically different.

This is a good place to for a summary of quantum particles and virtual quantum particles: Of all the quantum particles in our Universe described above, almost all have lifetimes of much less than a millionth of a second. Out of the 323 particles (Section 11) – 39 elementary particles and 284 hadrons plus an equal number of antiparticles, only 5 remain after a millionth of a second.

17. BOSONS MEDIATE INTERACTIONS BETWEEN QUANTUM PARTICLES.

Conservation of energy, momentum, charge, mass, and other key particle characteristics apply to every interaction (mediation) between particles.

The interactions between quantum particles are mediated by four **forces** (**bosons**) acting on (interacting with) the particles. The four BOSON forces are electromagnetic, strong nuclear (color), weak nuclear, and gravitational.

Each boson is its own anti-boson. Mass (M), Iso Spin (I), J Spin (J), Lifetime (LT), and charge (Q) are shown below for the bosons. Iso Spin and J Spin may be clockwise (+) or counterclockwise (-).

A. ELECTROMAGNETIC FORCE BOSON (PHOTON) (1).
ϒ **(Photon)**   M: 0 (less than 1×10^{-18})   I: 0.1   Spin (J): 1^- LT: ∞

B. WEAK NUCLEAR FORCE BOSONS (2). W^- is antiparticle for W^+.
W^+ M: 80,385   Spin (J): 1   LT: 10^{-25} Q: +1
W^- M: 80,185   Spin (J): 1   LT: 10^{-25} Q: -1
(Anti-W^+ boson)
Z^0 M: 91,187.6   Spin (J): 1   LT: 10^{-25} Q: 0

C. STRONG NUCLEAR (COLOR) FORCE BOSONS (GLUONS) (8).
G   M: 0   I = 0   Spin (J): 1^- LT: ∞   Q: 0

There are eight strong nuclear (color) force gluons.
Each gluon carries a combination of color and anti-color of quark colors red, blue, and green. To simplify, for most purposes, each gluon may be consider to carry a color and its anti-color such as red and anti-red. The discussion below provides additional information about the colors that each of the eight gluons carry.

D. GRAVITATIONAL FORCE BOSONS (3).
H^- is antiparticle for H^+. Graviton has not been observed.
H^0 Higgs 0   M: 126,000*   Spin (J): 0 (*)   LT: 10^{-24} (*)
H^+ Higgs +   M: 126,000*   Spin (J): 0 (*)   LT: 10^{-24} (*)
H^- Higgs -   M: 126,000*   Spin (J): 0 (*)   LT: 10^{-24} (*)
g (graviton)   M: 0 (less than 7×10^{-32})*   Spin (J): 2 (*)   LT: 10^{-24} (*)

*Gravitational force bosons were detected in July 1012 and are being extensively investigated at CERN.
The Higgs bosons and related gravitational fields are important in empty spacetime.

The Higgs boson was theorized as part of a search to find how the weak force W⁻ W⁺ Z⁰ bosons obtained mass. The Standard Model (SM) of particle physics was seen as deficient and of questionable validity until the existence of four Higgs fields was theorized to give mass to the weak force bosons and other particles along with the existence of Higgs bosons. The Higgs fields are responsible for the gravitational characteristics of the fabric of Space of our Universe and Space beyond our Universe.

The most common forces are the electromagnetic and gravitational forces. These are both "square-law" forces. The force F is attraction for between two particles (for gravity or electrical charges of opposite polarity, or repulsion (for electrical charges of the same polarity. The Force varies inversely with the square of the distance r between them: $F = C/r^2$ where C is a constant related to the strength of the Force. As the distance r between the particles approaches zero, the Force F approaches infinity asymptotically; as the distance r between the particles approaches infinity, the Force F approaches zero asymptotically. **See Sections 18 and 20**. The electromagnetic force is much stronger than the gravitational force – see below.

The strong nuclear (color) force is very different. It displays what is called "asymptotic freedom." It operates at short ranges also. When quarks within a hadron (meson or baryon) are close together they behave as free particles – free at least from the strong nuclear force. As distance between quarks in a meson or baryon increases, the strong nuclear force increases very rapidly and approaches infinity. If a "free" quark would stray and venture "outside" the hadron, the energy of the color force is strong enough to spontaneously create two quarks: an anti-quark to form a meson with the straying quark and another quark to replace it in the hadron from whence the straying quark came. Every interaction between particles has one or more mediator bosons. For electric, magnetic, and electromagnetic forces, it is the photon γ. For weak nuclear force, it is three bosons: W^+, W^-, and Z^0. For strong nuclear force, it is eight color-force gluons g. For gravitational force, it is theorized that there are the three Higgs bosons H^+, H^-, and H^0 and (perhaps) the as yet undetected graviton boson. The electromagnetic force, weak nuclear force, strong nuclear force, and gravitational force are thought to merge into a single force under extreme conditions of temperature and pressure.

18. DESCRIPTION AND RELATIVE STRENGTH OF BOSONS.

Under the conditions of most places in the Universe, there are wide differences in the relative strength of the four forces.

Starting with the weakest and ranking the forces in order by increasing relative strength:

Gravity	$= 10^{-42}$	H^+, H^-, H^0, and graviton bosons
Weak Nuclear	$= 10^{-13}$	W^+, W^-, and Z^0 bosons
Electromagnetic	$= 10^{-2}$	photons
Strong Nuclear Color	$= 10^1$	Eight strong nuclear (color) gluons

Each of the bosons is its own anti-boson. The W boson can act as a W^+ boson and also act as its anti-boson W^-. It is indicated as W^- when it is acting as an anti-boson. Bosons do not obey the Pauli Exclusion Principle.

Gluons (g) do not carry electric charge. Strong Nuclear force gluons carry "color" charge. Gluons can absorb and redistribute color charge. The W bosons can absorb and redistribute electric charge.

The Higgs boson is associated with the force of gravity. It was theoretical until July 2012 when it was reported found at the high-energy collider at CERN in Switzerland.

Photon bosons and strong nuclear (color) force gluon bosons are massless but carry their energy as their momentum. Weak force bosons and gravitational (Higgs) bosons have relatively large mass. The large masses of the weak force bosons give them very short de Broglie wavelengths. As a result, the weak force has a very short range of influence (10^{-18} meters) deep within an atom or proton which is much less than the radius of the proton (10^{-17}) centimeters. The weak force is thus short range rather than weak. The force of gravity is very weak but is infinite in range. The gravitational force of **extremely large masses, however, can be extremely large.**

The Higgs fields and Higgs bosons give mass to all particles that carry mass. The as yet undetected graviton boson is theorized to provide the force of gravitational attraction, but, the warpage of spacetime as described by Einstein's General Theory of Relativity explains gravitational phenomenon as the warpage of SpaceTime if the graviton is not found to exist.

The force of interaction between particles is generated by the exchange of bosons appropriate for the force being mediated between particles that carry the charge:
- Electromagnetic force photons for the electric charge (+ or -).
- Weak nuclear force bosons for weak isospin. (Weak isospin applies to leptons and quarks.)
- Strong nuclear force gluons for color charge of quarks, mesons, and baryons.
- Gravitational force bosons (and perhaps gravitons) for mass (force of gravity) for all particles that carry mass

Every interaction between particles has a mediator. Two protons or two electrons that have the same electric charge (+) "**know**" **that they are suppose to interact by repelling each other** because of the **CLOCKWISE** spin of mediator photon waves (bosons) are being passed between them pushing them apart. Similarly, a proton (+) and an electron (-) have different elect**rical charge and "know" that they are suppose to attract each other** because mediator **COUNTERCLOCKWISE** photon waves are pushing) them together.

While all this is going on, particles that have mass are being pushed together and being pulled toward other particles with mass by the Higgs fields and the warping of spacetime mediated by the Higgs bosons and possibly gravitons. In most cases, the photons win, as the electromagnetic force transmitted by photons is much stronger than gravitational attraction except where enormous mass is involved.

A very similar process takes place for the strong nuclear force interactions of the eight color charge gluons. For the weak nuclear force, W and Z^0 bosons do the job of mediating between particles.

The interactions between particles is determined by the forces acting on (interacting with) the particles. The four forces are electromagnetic, strong nuclear (color), weak nuclear, and gravitational ranking the forces in order by increasing relative strength.

19. HADRONS (MESONS, BOSONS, AND QUARK COMPOSITES) - DESCRIPTION.

There are several hundred hadrons of quark-composites) that have been discovered ETAC (detected Experimentally-To-A-Certainty). More will be discovered. For every hadron, an anti-hadron is possible. Anti-hadrons are made up of antiparticles. Most hadrons, except the neutron and proton are very short lived, much less than a millionth of a second. Then they decay further until a stable particle is formed.

Hadrons are composite particles made up of color-neutral combinations of quarks and bosons. The two types of hadrons are mesons and baryons. Quarks making up mesons and baryons are held together by Strong Nuclear (Color) Force Bosons (Gluons).

Hadrons also have corresponding anti-hadrons. For example, a proton consisting of three quarks has an anti-proton consisting of three counterpart anti-quarks. Putting a proton together with three quarks (uud) and an electron creates a hydrogen atom. Putting an anti-proton together with an anti-electron (a positron) and anti-bosons will produce anti-hydrogen.

This leads to the possibility of anti-Universes being created. An anti-Universe would annihilate itself and a universe like ours if contact was made.

20. EVERY QUANTUM PARTICLE IS PERTURBATION OF RELATED BOSON FORCE FIELD THAT STRETCHES TO INFINITY IN ALL DIRECTIONS - COULOMB'S LAW.

Sections 11 through 19 provide an overview of quantum particles. However, the understanding of quantum particles also requires broadening the perception of a quantum particle to treating it as a perturbation (disturbance) of an infinite force field over an infinite distance to understand how PARTICLES can act at infinite distances at infinite velocities.

A quantum particle itself does not travel - the perturbation of an infinite field which defines the particle travels. The behavior of a quantum particle cannot be described by considering it only as a particle. It must be viewed as an integral part of an infinite force field that stretches to infinity. Thus, the particle does not travel but the perturbation of ITS INFINITE FIELD travels FASTER THAN the velocity of light. The consideration of quantum particles in this manner, as part of a force field also ensures that every particle of a given kind is exactly identical to every particle of that type

Take a moment and write down Coulomb's Law for the interaction of two charged particles. When I ask students to do this, they all usually write down the following equation and I write it on the blackboard.

$F = (q_1 q_2)/r^2$ Where q_1 q_2 are two charged particles a distance r apart.

Then, I ask for all the students that think this is the correct equation to raise their hand: They all then will. Then I ask for all who think this is not the correct equation to raise their hands. I raise my hand. I enjoy the look of shock on the student's faces.

Here is the CORRECT FORMULA: $F = (q_1/r)(q_2/r)$ which is mathematically identical to the above formula.

Then the students are asked, "What are (q_1/r) and (q_2/r) ?
Answer this question before you read on.
Hopefully, a few students will answer, "Why those are equations for ELECTRIC FIELDS!!!   One electric field for each charge q_1 and q_2.

What is the extent of these two fields (q_1/r) and (q_2/r) ? Answer this question before you read on.

Answer: The value of each field in all directions will range from infinite at r = zero to zero at r = infinity!!!! Note that the electric field can be either negative or positive depending on the polarity of the charges. Like polarity charges will repel and unlike charges attract.

The same line of reasoning applies to gravity particles m of Newton's law of gravitational attraction between two masses m_1 and m_2 (Section 3).

$$F = (m_1/r)(m_2/r)$$

In the case of gravity, the gravitational fields also range from infinite at r = zero (0) to zero at r = infinity!

The value of a gravitational field (m/r) is, however, also modified by relative velocities and accelerations of the two masses according to Einstein's Special and General Theories of Relativity - that does not change the discussion here and is beyond the scope of this book. Refer to books on Einstein's Theories in the References for additional information.

Similar reasoning applies to interactions for the force of the other force bosons; however, those forces act only at short distances.

This infinite field characteristic of particles is the key to instant communication between particles over infinite distances.

Charge and mass (energy) are conserved in our Universe. If the fields of the two particles are perturbed one place, the change in the field will instantly appear everywhere else in the field even billions of km away or more apart.

The reader should note that electromagnetic waves are restricted to travel at the speed of light c (300 million meters per second) per Maxwell's equations. This restriction does not apply to waves of INFINITE fields between virtual particle pairs, entangled particle pairs, and probability waves which convey information instantly, faster than the speed of light as these are not only "particles" but also waves and infinite fields.

21. PARTICLE INTERACTIONS: THE HEART OF PARTICLE PHYSICS.

From the previous **Sections**, the pieces are in place to accomplish the objective of placing all the elementary particles of physics including the Higgs bosons and gravitons in their proper place in particle physics. This task is simplified because with some yet unproven exceptions, everything in the Universe and empty Space beyond is made of these particles. As was pointed out earlier, besides the electron and three neutrinos, out of the 285 "confirmed ETAC" hadrons that make up our Universe, only two are around more than a fraction of a microsecond: the proton (seemingly infinite lifetime) and neutron (14.67 minutes lifetime unless it is confined within an atom).

The proton and neutron are in turn are made up of quarks. The quarks are never found as particles as their boson is short range and immediately combines quarks into protons or neutrons or other hadrons. The rest of the hadrons aren't around very long as their lifetimes are only a minuscule part of a second. Add some electrons to protons and neutrons, and you can make all the atoms that are in the Universe. An electron is an electromagnetically charged lepton. Like the proton, it is long-lived (infinite), but the other two electromagnetically charged leptons, the *Muon* and Tau, decay in much less than a fraction of a second

Elementary particles interact with each other. The elementary particles interact to combine to produce hadrons; hadrons decay into stable particles; and Higgs bosons and gravitons impart mass to particles.

THREE WEAK NUCLEAR FORCE BOSONS W$^+$ W$^-$ Z^0 CONTROL PARTICLE INTERACTION AND DECAY.

ELECTROMAGNETIC FORCE BOSONS (PHOTONS) MEDIATE INTERACTIONS BETWEEN CHARGED PARTICLES.

22. WAYS THAT QUANTUM PARTICLES INTERACT.

When two particles approach each other, an interaction takes place. Particles can approach each other and then scatter (deflect away from or toward each other); particles can bind together and create hadrons; and particles particularly hadrons can decay into other particles. At the submicroscopic level of particle physics, conservation rules are the only way to tell exactly what goes on in any interaction between particles. Experimental physicists precisely measure all the characteristics of particles that are formed in particle interactions, usually when streams of extremely high-velocity electrons or streams of extremely high-velocity protons crash into each other or targets. Conserved characteristics of energy, momentum, charge, spin, and other parameters would be carefully measured.

Particle interactions are initiated in various ways. The latest method is two ultra-high-energy streams of protons are crashed together at the particle collider at CERN. Cosmic rays (mainly high-speed protons from the sun) bombard the earth and are also a readily available source of sporadic high energy electrons, protons, and other nuclei and particles at high altitude laboratories. After interaction(s), characteristics of the all particles that are left over and any new particles that are produced are precisely measured. Anomalies in the expected outcomes of interactions have in many cases resulted in discovery of new particles.

In some cases, the anomalies were due to errors in calculations or experimental technique. In many cases, theoreticians have predicted an outcome as the energy of the collider beams has been increased. This was the case with observation of the predicted Higgs boson at CERN in 2012. Observing the Higgs was a great victory for and confirmation of the Standard Model (SM) of Physics Particles. Had the Higgs not been found, the SM would have been proven to be incorrect and dozens and dozens of years of work by thousands and thousands of physicists would have had to be rethought and redone.

For example, electric charge (q) is a particle characteristic that must be conserved in interactions between particles. If high-energy streams of protons (with charge of +1) are crashed together, their three quarks with charges of +2/3, +2/3, and -1/3 interacting through strong-nuclear color force bosons, will always produce products with the have a total charge of +1 whether various other particles are mesons, baryons, electrons, positrons, photons, or whatever.

23. QUANTUM PARTICLE INTERACTION NOTATION.

The following notation is used in equations to describe the interaction of particles (facilitated by bosons as described in **section 17**). Charge, color and other characteristics are always conserved in interactions. These equations must always make sure that all conserved characteristics are properly accounted for on both sides of an equation.

23-A. CREATING ENERGY OUT OF MATTER. A quantum particle and its antiparticle can annihilate each other and create energy out of matter. For example, an electron (-) and an anti-electron (+) (positron) can interact through a photon boson and produce two uncharged photons. Note that the + and − electric charges cancel each other: **THE ELECTRON AND POSITRON INTERACTING THROUGH A PHOTON BOSON HAVE BEEN CONVERTED TO ENERGY PHOTONS (BOSONS)!!!!!** This is the mechanism (annihilation of virtual particles to create energy) that enables creation of a Universe!!!!!

$$e^- + e^+ + \gamma \rightarrow \gamma + \gamma$$

The reader can develop many more examples to create photons of heat energy (Section 2) using other particles and antiparticles (Section 2-A).
This created energy may be borrowed to create a Universe along with many, many billions and billions and trillions and many more interactions.

HOWEVER, the BALANCE OF EMPTY SPACE HAS BEEN DISRUPTED! Empty Space is now missing an electron e^-, a positron e^+, and two photon bosons γ. Ultimately; these must be paid back sooner or later perhaps at the end of the Universe along with all other "borrowed" particles and energy.

23-B. CREATING MATTER OUT OF ENERGY. Two photons can produce an electron and its antiparticle a positron: $\gamma + \gamma \rightarrow e^- + e^+$. As described above, the reader can determine many more examples of creating a particle and an antiparticle out of energy for every particle listed in Sections 16 and 17.

23-C. QUANTUM PARTICLE SCATTERING. Quantum particles can also interact by coming closely approaching each other (sometimes at very high velocities). In the below example, a photon interacts with an electron to scatter it (change its course) in a direction depending on the relative momentum of the two particles. The electron now carries all the momentum: $e^- + \gamma \rightarrow e^-$

23-D. CHARGED PARTICLES INTERACT THROUGH GLUONS.

Two up quarks u (charge +2/3 each) and one down quark d (charge -1/3) interact through a gluon g to produce a proton p(uud) (charge (+1):

$$u(+2/3) + u(+2/3) + d(-1/3) + g \rightarrow p(uud)(+1)$$

Or: more precisely, three quarks, each of a different color interact through a gluon to produce a colorless proton:

$$u^r + u^b + d^g + g \rightarrow pg$$

The gluon(s) remain within the proton to bind the three quarks together.

A neutron (udd) decays (mediated by a Z^0 boson) and changes to a proton (uud), electron, and anti-electron-neutrino. The proton weights less than the neutron.

$$n + Z^0 \rightarrow p^+ + e^- + \bar{\nu}_e$$

An electron and positron (mediated by a photon) create two quarks q (if the electron and positron energy is high enough) which then quickly (with binding of a gluon) combine to form a $q\bar{q}$ meson:

$$e^- + e^+ + \gamma \rightarrow q + \bar{q} + g \rightarrow q\bar{q}\, g$$

Showing the bosons (g, Z^0, γ) such as in the last four examples is included here for clarity. This level of detail is usually left to "Feynman diagrams." The W^+ W^- Z^0 weak force bosons are involved in some cases when necessary to maintain the conservation of charge from one side of an equation to the other as they can bring charge to an interaction or absorb charge to carry charge away.

The weak charge bosons then decay into other appropriately charged particles. Both charged weak force bosons (W^+ and W^-) being involved in a interaction at the same time is indicated as W meaning both W^+ and W^-. If the charge of a quark is not involved or does not change from one side of the equation to the other, then the Z^0 boson will mediate the interaction. π^+ Quarks: $u\bar{d}$ Q: 1   SCB: 0 0 0   M: 0139.57018   I: 1   J: 0
LT: 2.6033×10^{-8} Decay: $\mu^+ + v_u$

The π^+ meson consists of $u\bar{d}$ quarks (with a total charge Q of +1 (+2/3 and +1/3), with a LT of 2.6033×10^{-8} seconds. The π^+ meson decays into a anti-Muon μ^+ and Muon neutrino v_u (mediated by W bosons):

$$\pi^+ + W \rightarrow \mu^+ + v_u$$

The following is an extract from the "Review of Particle Physics" (RPP) which lists a quantity of 17 different kinds of N Baryons and includes the neutron n.

BARYON: N (Neutron) QTY: 17 SYMBOL: p, N^+, n, N^0 **QUARKS:** uud, udd
S = 0   **LT:** 880.1 and 10^{32} I = 1/2   J = 1/2, 3/2, 5/2, 7/2, 9/2, 11/2

There are 17 different kinds of **N** baryons. The neutron **n** is the most familiar **N** baryon. The **RPP** shows the n (neutron) Decay Modes as: p^+ e^- $\bar{v}_e$ 100 %.
A sub-listing indicates that a photon may also be present: p^+ e^- $\bar{v}_e$ Υ.

A neutron n (udd) decays (mediated by a Z^0 boson); the neutron decays into a proton (uud), electron, an anti-electron neutrino and sometimes a photon or two to carry any excess energy away.
$$n + Z^0 \rightarrow p^+ + e^- + \bar{v}_e + \Upsilon + \Upsilon$$

23-E. OVERVIEW OF PHYSICS OF A UNIVERSE.

By the time the readers get this far in this book, they are much more knowledgeable than they may realize about the physics of a Universe.

The goal in this book at this point is to have provided the information that Readers need to be able to understand the Physics of the creation and end of a Universe. At this point. The reader may wish to review the previous sections 2 through 23-D:

2: EINSTEIN'S EQUATION: Mass and energy are the same thing. A little mass makes a lot of energy:
2A. WORK, ENERGY, MASS, AND CONSERVATION LAWS.
2 and 3: VELOCITY OF LIGHT, ELECTROMAGNETIC SPECTRUM OF LIGHT.
5. GAMMA RAYS AND OTHER ELECTROMAGNETIC PHOTONS
7 AND 8: COSMOLOGICAL UNITS AND THE AGE AND DIAMETER OF THE UNIVERSE.
11: ELEMENTARY QUANTUM PARTICLES.
12. HEISENBERG'S UNCERTAINTY PRINCIPLE.
13. de BROGLIE WAVE CHARACTERISTICS OF ELEMENTARY QUANTUM PARTICLES.
14. QUANTUM PARTICLE PROBABILITY WAVES.
15. ENTANGLED QUANTUM PARTICLES.
16. VIRTUAL QUANTUM PARTICLES AND VIRTUAL QUANTUM ANTIPARTICLES.
17. BOSONS MEDIATE INTERACTIONS BETWEEN QUANTUM PARTICLES.
18. BOSONS - DESCRIPTION AND RELATIVE STRENGTH.
19. HADRONS (MESONS, BOSONS, AND QUARK-COMPOSITES) -DESCRIPTION.
20. EVERY QUANTUM PARTICLE IS PERTURBATION OF RELATED BOSON FORCE FIELD THAT STRETCHES TO INFINITY IN ALL DIRECTIONS: EXAMPLE: COULOMB'S LAW.
21. QUANTUM PARTICLE INTERACTIONS: THE HEART OF PARTICLE PHYSICS.
22. WAYS THAT QUANTUM PARTICLES INTERACT.
23. QUANTUM PARTICLE INTERACTION NOTATION.
23-A. CREATING ENERGY OUT OF MATTER.
23-B. CREATING MATTER OUT OF ENERGY.
23-C. QUANTUM PARTICLE SCATTERING.
23-D. CHARGED PARTICLES INTERACT THROUGH GLUONS.

24. ENERGY IS BORROWED FROM "EMPTY SPACE" TO CREATE VIRTUAL QUANTUM PARTICLES.

Virtual quantum particle pairs obtain energy by borrowing energy from infinite empty space.

The evidence of existence of these pairs of virtual quantum particles can be seen in many ways. For example, besides popping in and out of existence in "not-so-empty" empty space within our Universe and empty Space beyond Our Universe, virtual quantum particle quantum pairs also pop in and out of existence within an atom. Due to its negative charge, a virtual electron is closer to the positive protons in the nucleus: a virtual positron is closer to the orbiting electrons. If the atom is heated, the atom takes on energy and its electrons move to higher energy orbits. As it cools, electrons move to lower energy orbits and the atom emits energy by giving off photons (energy) at specific frequencies that make up the atom's spectrum. The virtual quantum electrons and positrons distort the spectrum, giving the atom what is called the "fine structure" of the spectrum of the atom. This fine structure is conclusive experimental proof of the existence of virtual quantum particle pairs.

Virtual Quantum Particles and Quantum Particles are exactly identical. Virtual Quantum Particle pairs come into existence by borrowing energy from empty space. Some may end up bound in atoms but most end up as free quantum particles until some are quickly annihilated by a close by antiparticles. Many quantum particles may live longer and are involved with particle interactions as described in Section 21, in the creation of atoms. "Short" and "long" lives are relative. For example, virtual quantum electron e^- and virtual quantum positron e^+ pair may pop into existence and annihilate each other into pure energy **(Section 22)** almost immediately. Or the pair may pop into existence, become separated, perhaps end up in an atom, or otherwise enjoy a much longer, perhaps an infinite life in "empty space." Other pairs may be created billions and billions of kilometers apart.

Long ago before our Universe (more than 13.8 billion years ago), there was (almost) infinite empty space. Infinite virtual quantum particle pairs existed there, and time there marched on.

24-A. COULOMB'S LAW DEMONSTRATES EXISTENCE OF VIRTUAL PARTICLES.

Coulomb's Law (Section 20) is: $\quad F = (q_1 q_2)/r^2$

Where q_1 and q_2 are two charged particles a distance r apart.

So should the two particles be attracted or repelled?

These two charged particles (such as electrons) communicate with the Universe and each other by each sending out electric fields of "streams of virtual photons" in all directions to infinity.

Depending on the polarity + or - of its charge, the virtual photon's "spin" is either is positive or negative.

If the charges of the two particles are the same, the photons spins will be the same, the two charged particles will each absorb the received photons and "kick" out photons toward the other particle so as to repel it away from the other particle. This is like the "kick" of a rifle opposite from the direction of the fired projectile. If the spins are different, the particles will each kick of photons away from the other particle to move it toward the other particle.

The net momentum and energy of all the particles is the same before and after the interaction just as it is when one billiard ball strikes another

Coulomb's law thus also illustrates the existence of virtual quantum particles (photons).

24-B. NEWTON'S EQUATIONS DEMONSTRATE EXISTENCE OF VIRTUAL PARTICLES.
Similar to Coulomb's Law as explained in Section 24-A for electrically charged particles, Newton's Law for gravitational attraction and Einstein's equations for relative time in his Special and General Relativity Theories **demonstrate the existence of virtual particles.** Here are some of the ramifications of these equations.

The "Force of Gravity" is only attractive in accordance with Newton's Law (Section 2). Gravitons are thought to propel objects toward each other in accordance with their relative masses in the same manner as photons propel electric charges toward (or away) from each other dependent on the magnitude of their charges. Gravitons, however, are theoretical and have never been found in nature or the laboratory.

24-C. TIME IS OF THE ESSENCE IN A UNIVERSE: KEY ASPECTS OF TIME IN A UNIVERSE AND EMPTY SPACE BEYOND.

Every object in a Universe (including People) always has a "Clock "with" it that keeps time regardless of the rate of clocks of other objects or particles. This clock is herein called the "On-Board Clock."

Every "On-Board Clock" just keeps on ticking at its same rate regardless of the relative rates of other On-Board Clocks moving at different relative velocities (Einstein's Special Relativity) or experiencing different accelerations including accelerations of lesser or greater gravitational fields (Einstein's General Relativity).

Clocks near massive objects such as large mountain run slower than clocks in a lesser gravitational field. If you are near a very massive object such the planet Jupiter, or a huge mountain, your clock will run slower.

Clocks moving faster than other clocks run slower than relatively stationary and slower moving clocks. On-Board clocks moving near the velocity of light have essentially (but not quite) stopped compared to slow moving clocks. Clocks moving at the speed of light in a vacuum have entirely stopped. It is impossible to travel faster than the velocity of light in a vacuum as time would have to run backwards.
For example, if you jump in a space ship and rush off to a distant star at near the speed of light, your clock will run much slower than clocks left behind. You many have aged 3 years when you return, but the people left behind may have aged thousands of years. This phenomenon has been proven by clocks placed in rapid orbit around the earth running slower that identical stationary clocks on the surface of the earth.

On-Board Clocks of cosmic particles approaching the Earth at near the velocity of light may last only microseconds as measured by their on-board clocks, but have much, much longer lives when measured by clocks on the Earth. **NO VELOCITY OF AN ON-Board CLOCK CAN EVER EXCEED THE VELOCITY OF LIGHT** in a vacuum.

24-D. CONVERSION OF MANY, MANY SMALL AMOUNTS OF MASS OF VIRTUAL QUANTUM PARTICLES INTO HUGE AMOUNT OF ENERGY CREATES UNIVERSE IN "EMPTY" SPACE.

NOTE: CREATION OF A UNIVERSE AND OTHER UNIVERSES REQUIRES THAT SPACE, TIME, AND VIRTUAL QUANTUM PARTICLES HAVE ALWAYS EXISTED EVERYWHERE. EMPTY SPACE IS NOT EMPTY BUT CONTAINS OUR UNIVERSE AND MANY OTHERS. SOME OTHER WRITERS BELIEVE THAT SPACE AND TIME DID NOT EXIST UNTIL OUR UNIVERSE WAS CREATED – A VIEWPOINT NOT SHARED HEREIN.

Infinite empty space is not empty. How could it be EMPTY? If it were empty, then its exact energy and momentum would be known everywhere there, and that would violate the "Uncertainty Principle" which forbids it (Section 12).

Where did the all the Energy in Empty Space come from to make BIG BANGS that Create Universes come from? A long time explanation in Cosmology is that the Einstein's General Theory of Relativity mathematics will provide the correct explanation, but that has not come about yet as the mathematics end with unsolvable "Singularities." The explanation herein is simpler:

Refer back to Section 2: Einstein's equation for the Equivalence of mass M and Energy E:

$$E = mc^2 = m\,(90{,}000{,}000{,}000\,000\,000)\ m^2/sec^2 = m\,(90\ \text{nonillion})\ m^2/sec^2$$

This equation means that a little bit of mass makes a lot of energy due to the factor c^2 multiplying mass m by the square of the velocity of light c^2.

The reader should realize by now that Virtual Quantum Particles are mass m and if enough particles of mass are converted to energy e (section 23), they would provide enough ENERGY to for a Big Bang to create a Universe. Section 23 gives an example of two virtual quantum particles annihilating and converting mass to energy but **the explanation applies to all particles (section 11).**

$$e^- + e^+ + \gamma \rightarrow \gamma + \gamma$$

This equation shows that a Virtual Quantum Electron e- and a Virtual Quantum Positron e+ will annihilate and convert to pure energy γ (photon).

24-D. CONVERSION OF MANY, MANY SMALL AMOUNTS OF MASS OF VIRTUAL QUANTUM PARTICLES INTO HUGE AMOUNT OF ENERGY CREATES UNIVERSE IN "EMPTY" SPACE (CONTINUED).

The energy to create the very tiny virtual quantum particles and convert them to pure energy has been borrowed from Infinite Empty Space. Ultimately this borrowed energy must be paid back to INFINITE EMPTY SPACE. However, in infinite empty space, there are enough virtual quantum particles available to be converted to provide enough pure energy in one place to make a Big Bang and create a Universe.

The remaining Sections explain the birth, existence, and death of our Universe, and other universes based on the Einstein's Equation $e = mc^2$ and its application to conversion of the mass of Virtual Quantum Particles to energy.

From Section 2-A: WORK, ENERGY, MASS, AND CONSERVATION LAWS:

Energy must be conserved in creation and end of a universe, so if energy is borrowed to create a universe, it must be repaid in full at the end of the universe and seek some evidence of this energy transaction.

There is evidence that this paying back of energy borrowed to create the Big Bang of our Universe is now taking place as explained in Section 34.

INFINITE NUMBER OF PAIRS OF VIRTUAL QUANTUM PARTICLE PAIRS BEYOND OUR UNIVERSE WILL PROVIDE ALL THE ENERGY REQUIRED TO CREATE BIG BANGS WHICH WILL CREATE UNIVERSES.

The previous section also explains why even a small amount of mass produces enormous amounts of energy, because of the multiplying factor c^2 (section 2) in Einstein's equation.

$$c^2 = 90,000,000,000,000,000 \text{ m}^2/\text{sec}^2$$

This is a very big number and it enables a large amount of heat energy to be obtained from a relatively small amount of mass.

Consider the mass of Virtual quantum Particles the electron e- and the positron e+. Each has an infinitesimal amount of mass of 9.109×10^{-31} kilograms. When they come together, they annihilate each other and convert their total mass 18.218×10^{-31} kilograms into pure energy (heat photons) as explained in the previous Section.

$$e\text{-} + e\text{+} + \gamma \rightarrow \gamma + \gamma$$

$$E = mc^2 = (18.218)\, 10^{-31} \text{ kilogram})\, (90 \times 10^{15} \text{ m}^2/\text{sec}^2) =$$

$$E = (1640)\, 10^{-16} \text{ kilogram m}^2/\text{sec}^2 = 1.640 \times 10^{-10} \text{ gram m}^2/\text{sec}^2$$

Thus tiny Virtual Quantum Electrons and tiny Virtual Quantum Positrons of total infinitesimally small mass 9.109×10^{-31} grams annihilate each other and produce a relatively enormous amount of pure energy in Empty Space of 1.640×10^{-10} gram m²/sec²

This mechanism applies to all Quantum Particles and enables creation of Universes.

25. BIG BANGS CREATE OUR UNIVERSE AND OTHERS OUT OF ENERGY BORROWED FROM VIRTUAL PARTICLE PAIRS IN "EMPTY SPACE" THAT ANNIHILATE INTO PURE ENERGY.

The fabric of empty space where our Universe was created was just like the fabric of empty space now within our Universe: teaming with virtual quantum particles (**Section 16**). Further, when future generations travel beyond our Universe (if that is ever possible), they will find empty space there is just like empty space of our Universe – **filled with virtual quantum particles. Verification of this assumption will hopefully soon be possible as the Space Exploration Program Voyager travels out of our Galaxy through empty space created between galaxies as galaxies move away from each other at an increasing rate.** It is expected in this book and postulated that on-board Voyager experiments will one day confirm the existence of virtual particles in this newly formed "empty space" between galaxies and verify the conclusions in this book. I am following up with the Voyager Program to encourage inclusion of this experiment.

One of the basic laws of all physics is that energy is conserved in any interaction: no energy – no interaction. If energy is "Borrowed" to cause an interaction, then the borrowed energy MUST BE EXACTLY PAID BACK (conserved) as the interaction unfolds. Thus, if there was nothing (no energy) before our Big Bang, there would have been no Big Bang! Thus, since there was a Big Bang, the energy to make our Big Bang had to be "BORROWED FROM SOMEWHERE" to which it must ultimately be repaid - perhaps repaid to **DARK ENERGY** (Section 34). The source of this borrowed energy to make a Big Bang is the infinite virtual quantum particle pairs of "Empty Space" that existed before our Universe and other Universes were created.   The borrowed energy ultimately must be repaid to "Empty Space" (Evidence of this repayment may be the Dark Energy that is causing our universe to expand.)

It also seems likely that **The Laws of Physics** everywhere will be the same as ours. Our Universe seems so carefully constructed that another type of universe even with only minor differences with physics laws will be unstable and swiftly collapse and disappear.

- Space and Time existed before our Universe in INFINITE EMPTY SPACE at the place where the Big Bang occurred and our Universe was created.

25-A. HOW MUCH BORROWED ENERGY IS REQUIRED TO MAKE A UNIVERSE?

Answer: Enough to create a Universe of a mass of 3×10^{56} kg (Section 3).

To Create a universe, the "virtual quantum particle pairs which randomly pop up (without outside influence) must "randomly pop up at a single point in Empty Space at the same time, in the amount necessary to create a Big Bang which will create a Universe. The following are two important aspects:

It takes 1.482×10^{107} pairs of virtual particles being created at exactly the same place at exactly the same time to create a universe.

The chance of this happening in our vicinity is infinitesimally small – not in our lifetime. But in an infinite amount of space in an infinite amount of time, universes will be randomly created an infinite number of times.

25-B. BIG BANG CREATED OUR UNIVERSE OUT OF ENERGY BORROWED FROM VIRTUAL QUANTUM PARTICLE PAIRS IN INFINITE EMPTY SPACE THAT EXISTED BEFORE OUR UNIVERSE EXISTED.

The Big Bang took place and our Universe was created in existing Infinite Empty Space and Time that contained Infinite virtual quantum particle pairs. "Energy" was created from the annihilation of virtual quantum particle pairs there into pure energy (Section 24-A. As more and more energy was added instantly from virtual quantum particles, GRAVITY INSTANTLY BECAME THE PRIMARY FORCE. Gravity instantaneously compressed the energy of all the Big Bang virtual particles into a smaller and smaller sphere as more and more the virtual particle energy was instantly added INCREASING GRAVITY. All the energy necessary to make the Big Bang from energy was instantly borrowed from virtual quantum particles of this empty space at the exact point and time of the Big Bang. Gravity concentrated this energy into an (almost) infinitely hot small sphere which then exploded with a Big Bang (Section 27).

Our Universe had a place and time to be born in - the empty space **TEAMING WITH VIRTUAL QUANTUM PARTICLE PAIRS** that existed before our Universe was created at the Big Bang. Our Universe is in infinite empty space where other Universes may exist but probably not nearby. Our Universe and these other universes would take umbrage at the statement, "TIME AND SPACE DID NOT EXIST BEFORE OUR BIG BANG."

There are probably googols (10^{100}) and googols and more googols of empty space outside our Universe and googols more Universes in existence. Or perhaps our Universe is the only one, or one of only a few. But EMPTY space is everywhere, within our Universe and beyond our Universe.

25-C. ENERGY TO MAKE A BIG BANG IS BORROWED ENERGY, CREATED FROM VIRTUAL QUANTUM PARTICLE PAIR ANNIHILATION IN EMPTY SPACE THAT EXISTED BEFORE OUR UNIVERSE, LEAVING BEHIND A UNIVERSE THAT MUST EVENTUALLY REPAY THIS BORROWED ENERGY.

This energy is now being paid back to Dark Energy (Section 34) causing our Universe to expand at an increasing rate , so as to return our Universe to Pre-Big-Bang condition as required by Hamilton's Principle and LaGrange's Equation.

The energy and rapid expansion of our new Universe during the Big Bang, at faster than the velocity of light, was an extremely rare improbable huge congregation of annihilated virtual particle pairs into energy at the exactly the same time and at exactly the same place where the Big Bang occurred. This huge instant annihilation of virtual particle pairs was a result of the VERY, VERY RARE random probability of all the required virtual particle annihilations to produce enough energy necessary to make Big Bang that created our Universe.

Left behind after borrowed energy made the Big Bang and created our Universe was the energy deficit in empty space that is now causing our Universe to expand at an increasing rate paying back the energy borrowed to make our Universe.

What about paying back the Borrowed Energy?
Answer: That process is going on now with Energy called "Dark Energy." Dark Energy (Section 32) is causing our Universe to expand at an increasing rate as galaxies move away from each other at an increasing rate.

26. CALCULATION OF QUANTITY OF VIRTUAL QUANTUM PARTICLE PAIRS THAT ANNIHILATED AT EXACT INSTANT AND PLACE OF BIG BANG.

Mass m of the Universe created by the Big Bang = 3×10^{56} kilograms.
Using Einstein's equation (Section 2) for the equivalence of mass and energy to calculate the amount of energy E it took to create the mass m of a Big Bang:

$E = mc^2$ where c is the velocity of light.

Energy E (Joules) required to create our Universe:

$E = (3 \times 10^{56} \text{ kg})(300 \times 10^8 \text{ m/sec})^2 = = 27 \times 10^{76}$ joules

Mass of a virtual electron and a proton pair = $(2)(9.109 \times 10^{-31})$ kg

NUMBER OF NEGATIVE (ELECTRON) AND POSITIVE (POSITRON) VIRTUAL PAIRS TO MAKE A BIG BANG:

$= 27 \times 10^{76} / 18.218 \ 109 \times 10^{-31}$ $= 1.482 \times 10^{107}$ pairs

NOTE: It is not the intent here to state that only electron/positron pairs would be involved in Big Bang energy, but to recognize that the annihilation of many other kinds of virtual particle pairs are also have been involved. The above result includes them as an approximation. All virtual pairs regardless of their particle types were spontaneously annihilated at exactly the same time and place to make our Big Bang.

26-A. DO OTHER UNIVERSES EXIST?

Answer: The calculations above show that creating a universe seems improbable and is likely infinitesimally rare. But in infinite Empty Space over infinite time before and after the present time, more Universes likely have been and will be created over infinite time and probably exist now somewhere. Someday PERHAPS SOON Our Decedents may learn how to make make contact with other Universes.

26-B. PAYING BACK ENERGY BORROWED TO MAKE G BANG.

That process of paying back the energy borrowed to create our Universe with a Big Bang is going on now with Energy called "Dark Energy." Dark Energy is Energy is causing our Universe to expand at an increasing rate until the borrowed energy of the Big Bang is paid back. When all the energy is paid back, our Universe will have entirely disappeared.

27. TIMELINE, TEMPERATURE, AND PRODUCTS OF BIG BANG THAT CREATED OUR UNIVERSE. (Adapted from Riordan)   Temperatures are in K (degrees Kelvin).

NOTE: Several conflicting "Big Bang Theories about How Our Universe Began" are being seriously studied. Refer to Hawkins's and other's letters in the Bibliography [Osborne].

1. 0 sec   ∞? K   Big Bang – Creation of Universe begins

Big Bang Sphere (BBS) has been created in infinite empty space. BBS consists of **ALMOST INFINITELY COMPRESSED QUANTUM PARTICLE ANNIHILATION ENERGY OF 27 X 10^{76} JOULES (3 x 10^{56} kilograms)** (Section 26), of about Planck diameter (10^{-33} cm), and almost infinite pressure and temperature. The compression and high temperature are caused by the force of gravity which now RULES SUPREME due to the **HUGE** energy of the Big Bang sphere.

2. 10^{-50} sec 10^{32} K   Plasma of particles.

3. 10^{-35} sec 10^{28} K   Inflation (also called inflaton)

A 10^{30} or more inflationary expansion of hot plasma in 10^{-35} sec at faster than the velocity of light. This plasma inflation was caused by pent-up pressure and energy of the Big Bang sphere expanding into the vacuum of infinite empty space. Inflation is created by all the energy (and mass) of the Universe from the huge energy of the Big Bang. Rapid expansion of the Universe then slows.

Our new Universe can initially expand faster than the velocity of light as it is expanding into empty Space just like our Universe today is expanding into empty space and relative motion of its galaxies moving away from each other does not exceed the velocity of light (Section 35).

4. 10^{-11} sec 10^{16} K   Transition

5. 10^{-7} sec 2×10^{13} K   Quarks and gluons form from plasma.

Cosmic Microwave background Radiation (CMBR) begins.

6. 10^{-5} sec 10^{12} K   Neutrons Form from quarks and gluons.

OUR UNIVERSE AND MULTI-UNIVERSES BEYOND

Quarks and gluons combine to form electrically neutral neutrons. Some electrically charged protons and electrons also form from quarks and gluons but mainly neutrons.

7. 10^2 sec 10^9 K   Neutrons Decay.

Neutrons decay into protons and electrons. Protons and electrons form clouds of hydrogen (77 percent), helium (23 percent), and lithium (trace). .

8. 10^5 yr 10^3 K   Stars form. Pressure causes hydrogen atom nuclei to combine in nuclear explosions and create helium.
9. 10^8 yr 10^2 K   Black holes form enabling Galaxies to form around them.
10. 10^9 yr 10^2 K   Solar Systems form
11. 10^{10} yr   2.7 K   Today

As Big Bang Cooled, Hydrogen Was Formed. Nuclear fusion of Hydrogen created the Elementary particles (photons, neutrinos, quarks, gluons, and electrons) and composite particles (neutrons and protons).

This extremely rapid inflation of our Universe was caused by the (almost) infinite pressure of the expanding Big Bang Sphere. The almost infinite pressure of the Big Bang and vacuum of empty space were thus working in conjunction with the enormous energy of the mass and energy of our Universe – a Universe out of virtual Quantum particles! It all took just 10^{-35} sec.

How big was the Big Bang? The Big Bang was just big enough and hot enough to get the process of inflation going. It took less than 10^{-50} sec.

At the instant of the Big Bang, temperatures were too extreme for anything to exist except formless plasma. In first micro-instants after the Big Bang, the plasma began to cool by radiating away energy as photons that are now seen on the Earth as the Cosmic Microwave Background Radiation (CMBR) described in **Section 31.**

27. TIMELINE, TEMPERATURE, AND PRODUCTS OF BIG BANG THAT CREATED OUR UNIVERSE (CONTINUED).

As the plasma continued to cool, quarks were able to form. At an instant later, quarks and gluons created electrically neutral particles: neutrons. The neutrons rapidly decayed into stable particles which have electrical charge: protons (+) and electrons (-), and uncharged neutrinos. Charged particles: protons (+) and electrons (-) were also produced at lower temperatures. The protons and electrons combined and in effect, converted the huge mass of the Big Bang into huge clouds of hydrogen.

The seemingly nondescript neutron that decays in about 15 minutes thus played a major role in the creation of our Universe by being responsible for 80 percent of its hydrogen. The neutron will later play a key role when stars turn into factories of all the elements (atoms) in our Universe.

The huge clouds of hydrogen were at the mercy of gravity, and condensed into denser and denser clouds creating protostars fusing hydrogen atoms into the more complex element helium and leading to the formation of stars.

The surface temperature (about 6000 K) of a star such as our sun is not high enough to support nuclear fusion to create helium. At its center, though, the temperature is about 15 million degrees K, high enough to fuse hydrogen into helium which gives off enormous fusion energy in the process. The surface temperature of most other stars ranges up to 60,000 K. The core temperature of the largest stars ranges up to 10^9 K.

After slowing down due to the effect of gravitational attraction of newly formed elements of the early Universe, our Universe has continued to expand and is now expanding at an accelerated rate due to the vacuum of empty space beyond our Universe. Our Universe will now continue to age and expand into cold oblivion.

Whoops!! What about anti-matter and antiparticles. As the plasma cooled after the Big Bang, both matter and anti-matter particles were created and quickly annihilated each other. Fortunately for every billion or so anti-matter particles that were created, one more matter particle than antimatter was created, or our Universe would never have been created. What was before the Big Bang? "Some scientists believe that Space and Time did not exist before the "Big Bang" created our Universe." That opinion is not shared here. Below is a different opinion substantiated by Black Hole Radiation?

28. TYPES OF STARS.

A solar mass is the mass of our sun (Msolar).
Stars range from about 0.2 solar masses to 300 solar masses.

PROTOSTAR. A huge cloud of stellar gas, mainly hydrogen, clumps and collapses due to gravitational attraction. Heat creates nuclear fusion changing hydrogen into helium and radiating neutrinos, photons, and gamma rays. This process accelerates over perhaps 100 million years and creates a star. There are billions and billions of galaxies, each with billions and billions of stars (**Section 7**).

1 SOLAR MASS (1.9885×10^{30} kg). Red Dwarf, Life Time: 10^{10} years. As it dies, it swells up to red giant, then, collapses to bright White Dwarf. Ends up cold cinders after about another 100 million years, Our Sun is about 5 billion years old. In another 5 billion years or so our Sun will swell up, engulf the Earth, shrink to a small bright white dwarf, run out of hydrogen, and then end up as cold cinders.

8 SOLAR MASSES. Becomes Supernova. Collapses electrons into protons and becomes neutron star of 1.4 solar masses, 10 Km radius, gravity = 10^{10} g.

MORE THAN 20 SOLAR MASSES. Explodes into supernova. Becomes neutron star of more than 2 solar masses. Collapses to almost infinite density and almost zero radius. Becomes a giant Black Hole that will suck in other stars and black holes.

DYING STARS ARE FACTORIES. All the atoms on Earth were once part of a dying star as were the atoms of the moons, planets, comets, and meteors in every galaxy.

Small stars that are dying create atoms of few protons – hydrogen, oxygen, carbon, and so forth (fortunately the ones needed for living things) but do not have temperatures high enough to create more complex atoms. Larger stars can create more complex atoms. But only the very rare largest stars have high enough temperatures when they die to create the most complex atoms such as gold and uranium.

29. DYING STARS ARE FACTORIES THAT CREATED ALL THE ATOMS IN A UNIVERSE FROM HYDROGEN CLOUDS.

As a star about the size of our sun dies (runs out of hydrogen), the pressure and heat of its fusion of hydrogen into helium are no longer able to keep it from collapsing. As it dies, it balloons up to a red giant and then collapses again; its core temperature rises greatly and helium in turn is fused into higher order elements such as boron, oxygen, and carbon (thank goodness, as all living things are made of carbon).

Mid-size stars have higher core temperatures, produce heaver elements, such as calcium, chlorine, sodium, and iron. Even including these stars, only about 26 of the natural 92 elements will have been produced.

Finally, the star collapses on itself or explodes, showering its galaxy with its remains, and for heaver stars, becoming supernova and neutron stars.

To produce the remaining 66 heavier elements, including the rare elements such as gold and uranium takes a star of about 20 solar masses or more. The core temperature of such star is 10 billion degrees K or more.

The range and types of elements being produced in the fiery factory of each star is different as the mass and core temperature of each star are different from those of other stars. A star's radiation (light) spectrum as detected by a telescope displays exactly what elements are being produced in each star by the colors and absorption lines of its spectrum.

30. BLACK HOLES AND QUASARS.

A dying star of more than 20 solar masses explodes into a supernova, and becomes neutron star of more than 2 solar masses. Then it collapses to almost infinite density and almost zero radius and creates a black hole that will suck in other stars and black holes.

- Diameter of a small black hole (of 5 stars) at the center of a galaxy:
    5 miles

- Diameter of a very massive black hole (of 5 million stars) at the center of a galaxy:
    5 million miles

- Temperature of 12-mile diameter black hole: 10^{11} K.
- Most galaxies, perhaps all, have a massive black hole at their center. Black holes may be essential and required by galaxies to be able to form around them.

- Quasars are very, very bright objects at long distances away that were formed with black holes in our early Universe. The amount of red shift of light from a quasar indicates how old it is. A quasar indicates that a galaxy and its associated black hole are being formed. As the black hole is formed, it accelerates matter into it which produces the bright light of the associated quasar.

- As reported in the magazine Nature, an especially brilliant quasar (the brightest and biggest quasar known) was recently discovered whose light was produced when the Universe was only 875 million years old. The black hole associated with this quasar contains the mass of 12 billion suns. The black hole associated with this quasar is among the biggest known even though it is among the oldest. About 40 quasars are known that developed within a billion years of the Big Bang.

- Our galaxy (Milky Way) has black hole of over 4 million solar masses at its center.
- Some galaxies have black holes at their center of several billion stars.

- Black holes radiate energy slowly. This is Hawking Radiation named after the author of the theoretical work, Stephen Hawking. Hawking Radiation is due to interaction of a black hole with virtual particles. If one of a pair of virtual particles (particle and antiparticle pair) is captured by the black hole releasing gravitational energy, its partner may escape as radiation. This radiation results in a loss of mass by the black hole, and eventually the black hole will entirely disappear as radiation.

- Radiation from black holes has not been detected experimentally.

- Black holes have limited life spans based on their mass: One star mass: 10^{67} years;

Galactic mass (billions of stars): 10^{97} years; multi-galactic mass: 10^{106} years.

31. COSMIC MICROWAVE BACKGROUND RADIATION (CMBR) IS EVIDENCE OF OUR BIG BANG.

Evidence of the explosion of the Big Bang that created Our Universe is heard as microwave noise and seen as snow on an old TV screen. This microwave radiation is called Cosmic Microwave Background Radiation (CMBR).

CMBR looks essentially the same in all directions. It is at the edge of the Universe. The radiation seen now occurred about 350,000 years after the Big Bang as the extreme temperatures of the Big Bang cooled to now very near absolute zero temperature (2.7 K). Since the CMBR looks almost the same from the Earth in all directions, it appears that the Earth is somewhere near the center of the Universe. At the Big Bang, every bit of space of our Universe was at one tiny location. This is why today that every galaxy seems (to itself) to be at the center of the Universe and why the CMBR looks the same in all directions to viewers at our galaxy and viewers at every other galaxy.

Cosmic Microwave Background Radiation (CMBR) was created very near the beginning of the Universe when it was too hot for matter to form (Section 27). The hot plasma of the Big Bang emitted light (microwave photons) now being detected from the fringes of the expanding Universe as the CMBR. As the Universe continued to rapidly expand, the hot plasma making the CMBR cooled and matter began to condense and form.
You can see CMBR as "snow" on over the air TV. Physicists were searching for the cause of the "noise" in their "radio telescope" and the hunt finally lead to the conclusion that the noise was radiation that took place shortly after the Big Bang.

The enormous energy of the Big Bang and its rapid inflation left an imprint on the CMBR that has been observed by scientists of the "Background Imaging of Cosmic Extragalactic Polarization (BICEPT2) Project." Their findings were announced in the _Los Angeles Times_ on March 18, 2014 in an article by Amina Kahn and will be published in _Nature._ This hard to detect imprint of gravitational waves on the CMBR changed the some of its polarization. Working at the South Pole to obtain clearer reception, the researchers were able to verify the change in polarization.

These findings about the CMBR at the South Pole support the theory that an inflationary Big Bang created our Universe. They also support the existence of gravitational waves.

32. DARK MATTER: 21 PERCENT OF OUR UNIVERSE; DARK ENERGY: 72 PERCENT; ORDINARY MATTER: 7 PERCENT.

Determining the composition of our expanding Universe is a problem. It is a big dark mysterious problem! The ordinary energy and matter that are detectable with telescopes and related instruments accounts for only about 7 percent of the amount needed to explain the motions of galaxies, stars, planets, and other objects of our Universe that can be "seen." The missing matter seems to be of two types: Dark (non-luminous and non-absorbing) Matter seemingly accounting for about 21 percent and Dark Energy seemingly accounting for about 72 percent of the missing matter. In March 2013, the Planck Satellite confirmed the values above, that dark energy was about 3 percent less than previously thought and ordinary matter is about 3 percent more.

There is currently extensive and intensive theoretical and experimental research regarding Dark Matter, Dark Energy, and Cosmology. According to the **2014 RPP, [Olsen, p. 368]**, "Experimental advances along these multiple axis could confirm today's relatively simple, but frustrating incomplete "standard model" of cosmology, or they could force yet another radical revision in our understanding of energy, or gravity, or the spacetime structure of our Universe. " The following discussion Dark Matter and Dark Energy extends the material of earlier Sections of this book to include the origin and composition of Dark Matter and the origin and operation of Dark Energy even though there is not universal agreement in these matters.

Beginning in 1933, cosmologists determined that a spinning cluster of galaxies did not seem to contain enough mass to keep the galaxies within it from flying out. Later, they determined as well, that a spinning galaxy did not seem to contain enough mass to keep the stars within it from flying out. No matter how hard or how long they looked, they couldn't detect, see, or find the necessary missing mass – so called Dark Matter.

The effect of a spinning galaxy cluster and a spinning galaxy is similar to a spinning disc with children riding on and going around with it. If a child is not very near the center, the centrifugal force of the spinning disc will throw the child out from the center and off the disc. With a spinning galaxy cluster, a spinning galaxy, and even the Earth spinning around the Sun, the force of attraction of gravity at the center must be enough to keep the galaxies, stars, and Earth from flying out and away.

33. COMPOSITION OF DARK MATTER AND ITS ORIGIN.

Dark matter exists. It has been mapped. Dark matter causes star light to bend as it passes dark matter on the way to telescopes.

The search is on for the composition of the missing dark matter (mass). Maps of dark matter have been made by various techniques showing that it extends and permeates out beyond the center of a galaxy. The bending of light by the gravity of dark matter has been used to make maps of dark matter. Using measurements of the variations of the cosmic microwave background radiation (CMBR) and of the special distribution of galaxies, finds a density of cold dark matter.

Candidates as listed in the **RPP** for dark matter include primordial Black Holes (created by burned out stars), WIMPs (Weakly Interacting Massive Particles) that are theoretical but as yet undetected, and Axions that are theoretically postulated by string theory but yet undetected. The detection of WIMPS requires underground laboratories to protect against background contamination by cosmic rays. The experiments searching for the composition of dark matter are detailed in the **RPP**. (See **Bibliography**.)

There is another origin and composition of Dark Energy theorized below suggested by review of the time line of the Big Bang (**Section 27**) and baryonic quark combinations (**Section 11**):

Shortly (10^{-7} seconds) after the Big Bang, quarks and gluons formed from the super-heated (10^{14} K) Big Bang Plasma (**Section 27**). This plasma created quarks and gluons, which in turn created quark composites. As far as is usually explained, the result was the creation of electrically-neutral neutrons. The neutrons decayed (after 880 seconds) and formed protons and electrons which then formed clouds of hydrogen atoms – the origin of stars. The protons will decay after 10^{32} seconds, but most all were immediately locked in atoms.

When quarks and gluons were formed 10^{-7} seconds after the Big Bang, the high temperatures (10^{14} K) would have created not only a simple quark combination like the neutron, but also various other neutral quark combinations of potentially 6, 9, 12, 15, etc quarks and maybe bunches in between. Most of these quark combinations would have decayed in much less than a microsecond, but some massive combinations, perhaps including the massive top quark, likely had a long lifetime like the proton, survived, and formed what is now seen and measured as dark matter.

The huge gravity of the surviving quark combinations might have been strong enough to clump them together as dark matter of a galaxy, and may have been necessary for galaxy formation.

Will these Dark Matter quark combinations if they exist ever be identified or created in the laboratory? Possibly, but it may be the temperatures and energies required are too extreme. However, it took over 40 years to finally create the Higgs boson ETAC in the collider at CERN….

34. DARK ENERGY (72 PERCENT OF OUR UNIVERSE) IS CAUSING OUR UNIVERSE TO EXPAND AT EVER-INCREASING RATE.

One way to find the total mass and energy in the Universe is to determine the rate of deceleration of the expansion of our Universe. Edwin Hubble had determined that the Universe was expanding in 1928. At that time, scientists expected that the Universe would gradually decelerate (stop expanding), and the force of gravity would ultimately cause all the objects and energy in the Universe to collapse "together" again as they were just before the Big Bang. Scientists set up experiments to find out for sure.

These experiments found that the Universe is expanding, but galaxies themselves are not expanding. Attractive gravity of a black hole at the center of each galaxy seems to hold each galaxy together. The distance between galaxies is expanding. Galaxies are like small coins pasted on a balloon. As the balloon inflates, the coins get farther apart, but the coins do not get any bigger.

The astonishing conclusion of various experiments in 1998 is that unexpectedly the expansion of our Universe expansion is not decelerating. Our Universe is not only continuing to expand, but the rate of expansion is accelerating. Universe expansion seems to have decelerated due to gravity of the Universe for the first 7 billion years or so after the Big Bang 13.8 billion years ago and now has unexpectedly accelerated at an ever-increasing rate.

The galaxies are not EXPANDING! The galaxies are moving away from each other in all directions. The Universe is EXPANDING. Galaxies in the Universe are moving away from each other. The rate of expansion of our Universe slowed at first, slowed by attractive gravity when the galaxies were closer together, but is now accelerating as the galaxies are moving further apart AT AN INCREASING RATE. **Galaxies do not increase in size. Empty Space between Galaxies is increasing causing our Universe to expand. This is like coins glued on a balloon. As the balloon is blown up, the balloon increases in size, the coins do not increase in size but move further apart.**

As described in the previous Section, a clue to what's going on is the energy of emptiness (empty space) that was outside the Big Bang sphere. Vacuum energy was created by the VACUUM OF EMPTY SPACE outside the Big Bang extreme energy (high temperature and high pressure) sphere inside it (**Section 25**).

Ultimately, the Big Bang Sphere could no longer hold the built-up and pent-up pressure and energy (**Section 27)** and it exploded with a **Big Bang.**

The unfurling (expansion) of our Universe is much like highly compressed gas in a tank **inside** a balloon suddenly being released into an enormous, a highly evacuated space. The highly compressed gas causes the balloon to be blown up quickly due to the vacuum pressure of the highly evacuated space. If the pressure of space was the same as that of the compressed gas, the balloon would not be blown up.)

At the earliest moments of the Big Bang, the enormous vacuum of empty space outside the expanding Universe **rapidly** inflated (blew up) the Big Bang sphere in all directions around it. This resulted in the creation of extremely high-temperature plasma to fill the expanding Universe. This subsequently formed all the gas clouds, galaxies, stars, planets, and other objects in our Universe.

Einstein's original equations implied that the Universe would expand. In disbelief, Einstein made changes. Then, when Hubble found that the Universe is expanding, Einstein realized that he had missed an opportunity of a lifetime to predict the expansion and reversed his changes. Now scientists realize that the Universe is not only expanding, it is expanding at an increasing rate. They realize as well that further changes to Einstein's equations are necessary to incorporate the effects of vacuum pressure beyond our Universe or whatever dark energy is causing the accelerating expansion of our Universe.

The hypothesis herein is that borrowed energy deficit left behind to create the Big Bang is the energy deficit called Dark Energy that is causing our Universe to expand at an ever increasing rate to pay back the borrowed energy Universe. Energy was borrowed to create the Big Bang and it is being paid back as our Universe continues to expand.

Ultimately, our Universe will continue to expand and all its mass will be converted to energy (Section 36) to repay the energy borrowed to create the Big Bang (Section 25).

The first spacecraft to ever leave our solar system, left in March 2013. It should not find anything untoward in its travels within our galaxy, the Milky Way. The real interest is when a space vehicle leaves our galaxy and travels in expanding space between galaxies. Things will get even more interesting when our ancestors are able to send a spacecraft beyond our Universe.

35. GALAXIES ARE MOVING AWAY FROM EACH OTHER AT INCREASING VELOCITY CREATING SPACE BETWEEN GALAXIES - BUT GALAXIES ARE NOT EXPANDING.

The Hubble Telescope has determined that our Universe is expanding, but galaxies are not expanding but are held in place by black holes. Only "Empty Space" between galaxies is expanding, and this newly created empty space will likely show exactly the same characteristics as empty space does everywhere now. The Voyager Space Program should verify the existence of virtual particles as its vehicles pass through "newly" created empty space between galaxies verifying experimentally the postulates made in this book.

SOON, no other galaxy will be visible from our galaxy and our Universe will be expanding faster than the velocity of light.

Impossible you say – nothing can move faster than the velocity of light. Yes, that is correct, but our Universe can expand faster than the velocity of light if you define terms properly.

Take a very big partially inflated balloon of 400 meters diameter to represent our universe.

Paste 400 coins around the diameter of the balloon, equally spaced 1 meter apart. The diameter of the balloon is 400 meters.
Now blow up the balloon continuously so the distance between coins increases 1 meter per second.
The diameter of the balloon is now increasing 400 meters per second.

Now blow up the balloon continuously so the distance between coins increases 1 million meters per second.

The diameter of the balloon is now increasing 400 million meters per second. This is faster than the velocity of light, but no coin is moving faster that velocity of light (300 million meters per second).

36. THE END OF OUR UNIVERSE.

Before our Universe was created, Empty Space was everywhere. It was just like empty space now in our Universe between galaxies. Empty Space before our Universe was populated with **virtual particles and virtual energy (Section 16)**. So is empty space in our Universe now. In fact, all the space in our Universe, even the space you are in right now - everything and many its atoms are populated with virtual particles (Lamb Effect). Without these virtual particles, a **Big Bang** could not have occurred. As a Voyager spacecraft leaves our Galaxy in the near future, hopefully, it will verify that virtual particles populate this "empty space" between galaxies just like all empty space in our galaxy.

It's interesting look to the end of the Universe for clues on this issue.

Near the end our Universe, all the stars will have expended their hydrogen fuel. Then, all matter in our Universe will eventually have been captured and "eaten" by Black Holes.

But, Black Holes radiate as discovered theoretically by Stephen Hawking in 1974. Within the Black Hole, particle and antiparticle pairs are produced. One of the pair of particles may become separated from the other, draw energy from the Black Hole to replace its partner, and escape from the Black hole as radiation, causing the Black hole to lose energy. The "former partner particle" or antiparticle left behind over time will repeat the process, combine into a particle, and be radiated.

Over a period of time (Section 30), a black hole will lose all its energy through this radiation and the Black Hole will disappear. Over a longer period of time, all Black Holes in our Universe will disappear in the same manner and **OUR UNIVERSE WILL BE NO MORE**! All the mass of our Universe will have been radiated away by Black Hole radiation and will have become part of "Empty Space" from which they came as part of paying back energy that were borrowed from infinite empty space to create our Big Bang (Section 25).

37. BIG BANGS CREATE OTHER UNIVERSES TOO, BUT NOT VERY OFTEN.

Our Universe - Was created in a Big Bang.
                 - Is 13.8 billion years old.
                 - Contains 4×10^{11} galaxies, 10^{23} stars (10^{12} stars per galaxy).
                 - Is expanding at an ever increasing rate.
                 - has a mass of 3×10^{56} kg.

Every Galaxy in a Universe has one or more Black Holes at its center. A Black Hole is essential to formation of a Galaxy. Galaxies are moving away from each other at an accelerating rate. Galaxies are not expanding. A Galaxy is prevented from expanding by Black Hole(s) at its center. Space between galaxies is expanding at an accelerating rate. This expanding space between Galaxies causes our Universe to expand at an accelerating rate.

A Universe is essentially homogeneous and isotropic: All places in the Universe are equivalent. Every place in the Universe is at the center of the Universe. No matter how far and where you travel in the Universe, you will find the galaxy you travel to will be at the center of the Universe. You can never travel out of our Universe, as you will always be at the center of our Universe!

Light from a distant place in the Universe is red-shifted according to its velocity, showing the Universe is expanding at an accelerating rate.

All the stars in the Universe will eventually burn out and either collapse into a Black Hole or be swallowed up by another Black Hole. Every Black Hole in Our Universe will ultimately radiate away by Hawking Radiation. Our Universe will then have completely disappeared and only the empty space will remain.

Our Universe is like a house of cards. Change even one simple aspect of our Universe and our Universe would completely collapse into nothing. Every aspect of our Universe is closely related to every other aspect. Other Universes that might operate differently than ours, probably could not be created, probably could not exist, and probably could not function.

Any budding universe which might be different probably would collapse like a house of cards into an empty space-time of virtual particles waiting for a proper Universe to be created. For example, where would Coulomb's law be without photon spin; where would Einstein's theory of gravity and space-time be without even one of the four Higgs bosons?

Even to understand Coulomb's Law of Interaction of electrically charged particles (Section 20) requires touching on almost every major topic of explanation of our Universe. Every major topic of our Universe is closely related to and dependent on every other topic.

The physics of our Universe seems to be so interdependent that one small change may send everything crashing to destruction. It seems likely that all Universes will be quite like ours from a physics standpoint. Otherwise, if we were to encounter a different universe, both universes might be destroyed. If another universe is created with different physics, then positive and negative matter would may likely be created equally and destroy each other at their Big Bang, unlike our Universe which had a slight preference for positive matter.

There are perhaps googols (10^{100}) and googols and more googols of empty space outside our Universe and googols more Universes in existence. Or perhaps our Universe is the only one, or one of only a few. But space is everywhere, within our Universe and beyond out Universe.

Finally a consistent explanation of the birth, existence, and death of our Universe as explained in this book requires the assumption of virtual particles before, during, and after our Universe.

When will the next Universe be created?

Answer: That's like asking how long it will take to throw $10^{1000000000}$ dice at exactly the same time and have all of them all come up boxcars. It could be anytime or billions of years, or more. But over infinity it will happen, perhaps many times. The probability can be computed.

Do other Universes exist?

Answer: In our infinite Empty Space over the infinite time before and after the present, more Universes likely have been created and some besides ours may exist now somewhere. We may learn how to make "Wormholes" and make contact with other Universes in the future.

EPILOGUE: SPACE AND TIME EXISTED BEFORE OUR BIG BANG AND WILL CONTINUE TO EXIST AFTER OUR UNIVERSE CEASES TO EXIST.

This postulate is controversial! Not all physicists agree. Nevertheless, it seems likely to this writer that our Universe had to have a place and time to be born and obtain the energy necessary to make a big bang. Our Universe did not exist before the Big Bang so there were no clocks timing events in our Universe until the Big Bang. Nevertheless, time and empty space did exist everywhere before our Universe was created. They do exist now outside and beyond our Universe. Just as in our Universe, identical clocks in other universes will run either faster than other identical clocks depending on relative velocities and acceleration.

This book also postulates that Empty Space outside of and before our Universe was full of "Virtual Particle pairs" just like empty space everywhere in our Universe was and is full of virtual particle pairs today! Anytime the term "empty space" is used in this book, the Reader should assume it is teaming with virtual particle pairs.

Space beyond our Universe was, is, and will be just like empty space in our Universe, including the "empty space" in our Universe that is being "created" now as galaxies within in our Universe move away from each other at an ever increasing rate.

Time and space within our Universe began with the Big Bang at a time and place that existed long before our Universe was created. Our Universe will disappear into this space after our Universe has ceased to exist by paying back all the energy borrowed to create our universe causing our universe to expand at an increasing rate.

OUR UNIVERSE AND MULTI-UNIVERSES BEYOND

It's hard to determine what existed before Our Universe was created as there is no evidence about anything until the Big Bang. But if we can figure out what exists after all traces of our Universe disappear, then we will can extrapolate to conditions before our Big Bag and know exactly what existed before Our Universe existed. Fortunately, thanks to the continuing Genius of Stephen Hawking, we are now able to observe how Our Universe will end and cease to exist.

Our Universe is expanding at an ever increasing rate as detected by the Hubble Telescope giving us hard data about how our Universe will end. All galaxies are moving away from each other at an ever increasing rate creating empty space between them. The characteristics of this empty space between galaxies are the same as empty space within our galaxy, including hosting virtual particles. The existence of virtual particles between galaxies hopefully can be confirmed by experiments of Voyager probes, which are now moving between nearby solar systems. Cosmological studies have shown that all stars will eventually burn out and create Black Holes or be eaten by other Black Holes. There will come a time when only Black Holes will remain of our Expanding Universe.

If we know conditions after our Universe ceases to exist, then those conditions will be the same as existed everywhere outside Our Universe before the Big Bang of Our Universe. Steven Hawking determined that Black Holes create virtual particles and then radiate virtual particles away. Ultimately, all black Holes in our Universe will radiate away in virtual particles paying back energy that was borrowed to create the Big bang. This payback of energy is now causing our Universe to expand at an ever increasing rate. Ultimately all the energy of the Big Bang will be radiated away as virtual particles from dying Black Holes fully repaying the borrowed energy used to make the Big Bang that created our Universe. Empty space beyond our Universe thus turns out to be just like empty space in our Universe. This was the last piece of the puzzle for me in understanding our Universe (See **Foreword**).

EPILOGUE: (CONTINUED).
SPACE AND TIME EXISTED BEFORE OUR BIG BANG AND WILL CONTINUE TO EXIST AFTER OUR UNIVERSE CEASES TO EXIST

There are many theories of our Universe. The one in this book is the only one that I feel adequately reflects the current experimental and observational data and information. I believe that it will pass the Test of Time and will continue to be relevant long after other theories have been forgotten.

Our Universe is a very strange place. The Hubble Telescope has shown that "every Galaxy" is at the center of our Universe and no matter where you travel in our Universe, you will always be at the "center of our Universe."

Our Universe is rapidly expanding as galaxies are moving away from each other at an increasing rate. Eventually, no other galaxy will be able to be seen from any other galaxy.

It's interesting to read about the various shapes of our Universe that mathematical physicists have proposed. They all agree our Universe is "Curved Somehow."

It also seems reasonable that creation of a universe is a very, very rare event. Nevertheless in infinite empty space, an infinite number of universes will be created in an infinite amount of time.

THANK GOODNESS, WE HITCHED A RIDE ON THIS ONE!!!!!

JGB

BIBLIOGRAPHY.

The material listed below was reviewed and is recommended for further reading particularly from a historical viewpoint which was essentially ignored herein. Much material is similar between references and it is difficult to single out and cite any one reference in any specific area except as provided by the titles or as noted below.

In cases of conflict between sources the *Review of Particle Physics* (*RPP*) prevailed. Citations in the text to **BIBLIOGRAPHY** listings are enclosed in brackets [ ].

[1] Baggott, Jim, *The Quantum Story,* Oxford University Press (2011).

[2] Baggott, Jim, *Higgs,* Oxford University Press (2012).
A historical summary of the search for the Higgs boson by a participant.

Bean, John Gilbert, The Universe, Space, and Beyond, Create Space .com (10/13/16)

Beringer, J. et al, *Review of Particle Physics* **(Particle Data Group), Physical Review D, Vol. 86 010001** A 1525 page definitive summary of the latest research and experimentation in particle physics. An abbreviated pocket booklet is also published; order it at: http://pdg.1b1.gov **(2012).** See also listing for Olive, K.A.

Bohm, David, *Quantum Theory,* **Dover Publications, Inc. (1951).**
A qualitative and physical presentation of fundamentals followed by considerable mathematical detail.

Collier, Peter, *A Most Incomprehensive Thing: Notes Toward a Very Gentle Introduction to the Mathematics of Relativity* **(2012). (Available at Amazon.com)**

Condon, E. U. and Odishaw, Hugh, *Handbook of Physics,* **McGraw Hill (1967).**

BIOGRAPHY (CONTINUED).

Freeman, Morgan, **"Through the Wormhole (Video),** "Is Gravity an illusion?" Science Channel.

Feynman, Richard P., *QED the Strange Theory of Light and Matter*, Princeton University Press (2006).
Richard Feynman is developer of Feynman diagrams of particle interactions.

Gilmore, Robert, **Lie Groups, Lie Algebras, and Some of Their Applications,** Dover (2002).

[1] **Greene, Brian**, *The Hidden Reality,* Vintage Books (2011).
[2] **Greene, Brian**, *The Fabric of the Cosmos*, Vintage Books (2004).
[3] **Greene, Brian**, *The Elegant Universe,* Vintage Books (2000).

Griffiths, David, *Introduction to Elementary Particles*, Wiley-VCH (2010).
A textbook on the mathematics of Particle Physics. Chapter 10 "Gauge Theory) gives the mathematical derivation of the Higgs boson from the Lagrangian.

[1] **Hawking, Stephen**, *A Brief History of Time* (2004).
[2] **Hawking, Stephen and Mlodinlow, Leonard**, **THE GRAND DESIGN** (2008).
[3] **Hawking, Stephen**, *Curiosity – Did God Create the Universe,* Discovery Productions, DiscSc Channel, (Video recorded 8/7/2011).

Krass, Lawrence M. *A Universe from Nothing*, Free Press (2012).

Kay, David C, *Tensor Calculus,* McGraw-Hill (1988

Leiber, Lillian R, *The Einstein Theory of Relativity,* Paul Dry Books (2008)
General Relativity using tensor notation and calculus.

Lipschutz, Seymour and Lipson, Marc, *Linear Algebra,* McGraw-Hill (2002).

[1] McMahon, David, *Relativity*, McGraw-Hill (2006).
[2] McMahon, David, *String Theory*, McGraw-Hill (2009)

Olive, K. A. et al, *Review of Particle Physics* (Particle Data Group) (2014)

Palen, Stacy, *Astronomy,* McGraw-Hill (2002), D. Van Nostrand Co., Inc. (1955)

Riordan, Michael and Zajc, William A, *The First Few Microseconds*, Scientific American (May 2006)

Robertson, John, *Geometrical and Physical Optics,* D. Van Nostrand Co., Inc. (1955)

Schumm, Bruce A, *Deep Down Things,* The John Hopkins University Press (2004).

Stannard, Russell, Relativity*, a Very Short Introduction,* Oxford University Press (2008).

Sternheim, Morton M, and Kane, Joseph W, G*eneral Physics,* John Wiley and Sons (1986)

ALPHABETICAL INDEX

ABOUT THE AUTHOR

BIBLIOGRAPHY

25. BIG BANG CREATED OUR UNIVERSE OUT ENERGY BORROWED FROM VIRTUAL PARTICLE PAIRS THAT EXISTED BEFORE OUR UNIVERSE EXISTED

26-B. BIG BANG, PAYING BACK ENERGY BORROWED TO MAKE BIG BANG.

27. BIG BANG, TIMELINE, TEMPERATURE, AND PRODUCTS OF THAT CREATED OUR UNIVERSE

30. BLACK HOLES AND QUASARS

37. BIG BANGS CREATE OTHER UNIVERSES TOO, BUT NOT VERY OFTEN.

18. BOSONS DESCRIPTION AND RELATIVE STRENGTH

17. BOSONS MEDIATE INTERACTIONS BETWEEN QUANTUM PARTICLES

23-D. CHARGED PARTICLES INTERACT THROUGH GLUONS.

31. COSMIC MICROWAVE BACKGROUND RADIATION (CMBR) IS EVIDENCE OF OUR BIG BANG

6. COSMIC RAYS

24-D. CONVERSION OF MANY, MANY SMALL AMOUNTS OF MASS OF VIRTUAL QUANTUM PARTICLES INTO HUGE AMOUNT OF ENERGY CREATES UNIVERSE IN "EMPTY" SPACE.

7. COSMOLOGICAL UNITS

20. COULOMB'S LAW

24-A. COULOMB'S LAW DEMONSTRATES EXISTENCE OF VIRTUAL PARTICLES.

34. DARK ENERGY (72 PERCENT OF OUR UNIVERSE) IS CAUSING OUR UNIVERSE TO EXPAND AT EVER-INCREASING RATE

33. DARK MATTER COMPOSITION OF AND ITS ORIGIN

32. DARK MATTER: 21 PERCENT OF OUR UNIVERSE; DARK ENERGY: 72 PERCENT; ORDINARY MATTER: 7 PERCENT

13. de BROGLIE WAVE CHARACTERISTICS OF ELEMENTARY QUANTUM PARTICLES

DEDICATION.

5. ELECTROMAGNETIC SPECTRUM OF PHOTONS

EPILOGUE   SPACE AND TIME EXISTED BEFORE OUR BIG BANG AND WILL CONTINUE TO EXIST AFTER OUR UNIVERSE CEASES TO EXIST.

11. ELEMENTARY QUANTUM PARTICLES

24. EMPTY SPACE WITHIN OUR UNIVERSE AND OUTSIDE OUR UNIVERSE

23-A. ENERGY, CREATING OUT OF MATTER.

25-A. ENERGY, HOW MUCH BORROWED ENERGY IS REQUIRED TO MAKE A UNIVERSE?

25-B. ENERGY, BIG BANG CREATED OUR UNIVERSE OUT OF ENERGY BORROWED FROM VIRTUAL QUANTUM PARTICLE PAIRS IN INFINITE EMPTY SPACE THAT EXISTED BEFORE OUR UNIVERSE EXISTED.

25-C. ENERGY TO MAKE A BIG BANG IS BORROWED ENERGY, CREATED FROM VIRTUAL QUANTUM PARTICLE PAIR ANNIHILATION IN EMPTY SPACE THAT EXISTED BEFORE OUR UNIVERSE,

15. ENTANGLED PARTICLES

FOREWORD.

35. GALAXIES ARE MOVING AWAY FROM EACH OTHER AT INCREASING VELOCITY

35. GALAXIES, CREATING SPACE BETWEEN - GALAXIES ARE NOT EXPANDING

19. HADRONS (MESONS, BOSONS, AND QUARK-COMPOSITES) - DESCRIPTION

12. HEISENBERG'S UNCERTAINTY PRINCIPLE

23. INTERACTION NOTATION

9, 10. TEMPERATURE SCALES, OF VARIOUS PHENOMENA

4. LIGHT, VELOCITY OF

2. MASS AND ENERGY, EINSTEIN'S EQUATION FOR EQUIVALENCE

3. MASS (IN KILOGRAMS) OF VARIOUS OBJECTS

23-B. MATTER, CREATING MATTER OUT OF ENERGY.

24-B. NEWTON'S EQUATIONS DEMONSTRATE EXISTENCE OF VIRTUAL PARTICLES.

2. NEWTON'S LAW OF GRAVITY

22. PARTICLES INTERACT, WAYS THAT

OUR UNIVERSE AND MULTI-UNIVERSES BEYOND

ALPHABETICAL INDEX

21. PARTICLE INTERACTIONS: HEART OF PARTICLE PHYSICS

PREFACE

20. QUANTUM PARTICLE IS PERTURBATION OF RELATED BOSON FORCE FIELD THAT STRETCHES TO INFINITY IN ALL DIRECTIONS: COULOMB'S LAW

14. QUANTUM PARTICLE PROBABILITY WAVES.

23-C. QUANTUM PARTICLE SCATTERING.

11-B. QUANTUM PARTICLES (SECTION 11) ARE INFINITE PROBABILITY WAVES THAT INSTANTLY SPREAD AND EXTEND THROUGHOUT ENTIRE UNIVERSE.

16. VIRTUAL QUANTUM PARTICLES AND VIRTUAL ANTIPARTICLES

28. STARS, TYPES OF

29. STARS ARE FACTORIES THAT CREATED ALL THE ATOMS IN 0UR UNIVERSE FROM HYDROGEN CLOUDS

26. VIRTUAL PARTICLE PAIRS, CALCULATION OF QUANTITY OF THAT ANNIHILATED AT EXACT INSTANT AND PLACE OF BIG BANG

16. VIRTUAL PARTICLES AND VIRTUAL ANTIPARTICLES.

25. VIRTUAL PARTICLE PAIRS, BIG BANG CREATED OUR UNIVERSE OUT ENERGY BORROWED FROM

2-A. WORK, ENERGY, MASS, AND CONSERVATION LAWS

36. UNIVERSE, END OF

8. UNIVERSE, AGE AND DIAMETER OF

8. UNIVERSE, DIAMETER OF

11-A. UNIVERSES, ARE UNIVERSES MADE OF ANTI-ELECTRONS AND ANTI-PROTONS POSSIBLE?

26-A. UNIVERSES, DO OTHER UNIVERSES EXIST?

1. UNIVERSE, TIMELINE AND INFORMATION ABOUT

www.ingramcontent.com/pod-product-compliance
Lightning Source LLC
Chambersburg PA
CBHW082352220526
45470CB00008B/2724